T0254684

Compact Textbooks in Mathematics

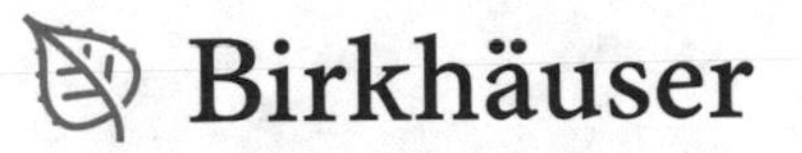

Compact Textbooks in Mathematics

This textbook series presents concise introductions to current topics in mathematics and mainly addresses advanced undergraduates and master students. The concept is to offer small books covering subject matter equivalent to 2- or 3-hour lectures or seminars which are also suitable for self-study. The books provide students and teachers with new perspectives and novel approaches. They feature examples and exercises to illustrate key concepts and applications of the theoretical contents. The series also includes textbooks specifically speaking to the needs of students from other disciplines such as physics, computer science, engineering, life sciences, finance.

- **compact:** small books presenting the relevant knowledge
- **learning made easy:** examples and exercises illustrate the application of the contents
- **useful for lecturers:** each title can serve as basis and guideline for a semester course/lecture/seminar of 2–3 hours per week.

More information about this series at: http://www.springer.com/series/11225

Alessandro Fonda

The Kurzweil-Henstock Integral for Undergraduates

A Promenade Along the Marvelous Theory of Integration

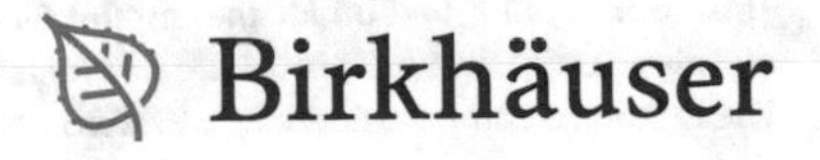

Alessandro Fonda
Dipartimento di Matematica e Geoscienze
Università degli Studi di Trieste
Trieste, Italy

ISSN 2296-4568 ISSN 2296-455X (electronic)
Compact Textbooks in Mathematics
ISBN 978-3-319-95320-5 ISBN 978-3-319-95321-2 (eBook)
https://doi.org/10.1007/978-3-319-95321-2

Library of Congress Control Number: 2018949935

Mathematics Subject Classification (2010): 26A39, 26A42, 26A63, 26A66, 26B15, 26B20, 28A75

This book is published under the imprint Birkhäuser, www.birkhauser-science.com by the registered company Springer Nature Switzerland AG
The registered company address is: Gewerbestrasse 11, 6330 Cham, Switzerland

The author with Jaroslav Kurzweil in September 2008

To Sofia, Marcello, and Elisa

Introduction

This book is the outcome of the beginners' courses held over the past few years for my undergraduate students. The aim was to provide them with a general and sufficiently easy to grasp theory of the integral. The integral in question is indeed more general than Lebesgue's in $\mathbb{R}^N$, but its construction is rather simple, since it makes use of Riemann sums which, being geometrically viewable, are easily understandable.

This approach to the theory of the integral was developed independently by Jaroslav Kurzweil and Ralph Henstock since 1957 (cf. [5, 8]). A number of books are now available [1, 4, 6, 7, 9–13, 15–19, 21]. However, I feel that most of these monographs are addressed to an expert reader, rather than to a beginner student. This is why I wanted to maintain here the exposition at a very didactical level, trying to avoid as much as possible unnecessary technicalities.

The book is divided into three main chapters and five appendices, which I now briefly describe, mainly as a guide for the lecturer.

The first chapter outlines the theory for functions of one real variable. I have done my best to keep the explanation as simple as can be, following as far as possible the lines of the theory of the Riemann integral. However, there are some interesting peculiarities.

- The Fundamental Theorem of differential and integral calculus is very general and natural: one only has to assume the given function to be primitivable, i.e., to be the derivative of a differentiable function. The proof is simple and clearly shows the link between differentiability and integrability.
- The generalized integral, on a bounded but not compact interval, is indeed a standard integral: in fact, Hake's theorem shows that a function having a generalized integral on such an interval can be extended to a function which is integrable in the standard sense on the closure of its domain.
- Integrable functions according to Lebesgue are those functions which are integrable and whose absolute value is integrable, too.

In the second chapter, the theory is extended to real functions of several real variables. No difficulties are encountered while considering functions defined on rectangles. When the functions are defined on more general domains, however, an obstacle arises concerning the property of additivity on subdomains. It is then necessary to limit one's attention to functions which are integrable according to Lebesgue, after having introduced the concept of measurable set. On the other hand, for the Fubini Reduction Theorem there is no need to deal with Lebesgue integrable functions. It has a rather technical but conceptually simple proof, which only makes use of the

Kurzweil–Henstock definition. In the Theorem on the Change of Variables in the integral, once again complications may arise (see, e.g., [2]), so that I again decided to limit the discussion only to functions which are integrable according to Lebesgue. The same goes for functions which are defined on unbounded sets. These difficulties are intrinsic, not only at an expository level, and research on some of these issues is still being carried out.

The third chapter illustrates the theory of differential forms. The aim is to prove the classical theorems carrying the name of Stokes, and Poincaré's theorem on exact differential forms. Dimension 3 has been considered closely: indeed, the theorems by Stokes–Cartan and Poincaré are proved in this chapter only in this case, and the reader is referred to Appendix B for the general proof. Also, I opted to discuss only the theory for M-surfaces, without generalizing and extending it to more complex geometrical objects (see however Appendix C). In some parts of this chapter, the regularity assumptions could be weakened, but I did not want to enter into a topic touching a still ongoing research.

In Appendix A, the basic facts about differential calculus in $\mathbb{R}^N$ are reviewed.

In Appendix B, the theorems by Stokes–Cartan and Poincaré are proved. The proofs are rather technical but do not present great conceptual difficulties.

In Appendix C, one can find a brief introduction to the theory of differentiable manifolds, with particular emphasis on the corresponding version of the Stokes–Cartan theorem. I did not want to deal with this argument extensively, and the proofs are only sketched. For a more complete treatment, we refer to [20].

In Appendix D, one of the most surprising results of modern mathematics is reported, the so-called Banach–Tarski paradox. It states that a three-dimensional ball can be divided into a certain number of subsets which, after some well-chosen rotations and translations, finally give two identical copies of the starting ball. Why reporting on this in a book about integration? Well, the Banach–Tarski paradox shows the existence of sets which are not measurable (a rotation and a translation maintain the measure of a set, provided this set is measurable!), and it does this in a very spectacular way.

Appendix E entails a short historical note on the evolution of the concept of integral. This note is by no means complete. The aim is to give an idea of the role played by the Riemann sums in the different stages of the history of the integral.

Note A preliminary version of this book was published in Italian under the title *Lezioni sulla teoria dell'integrale*. It has been revised here, extending and improving most of the arguments.

Contents

1 Functions of One Real Variable

Alessandro Fonda

A. Fonda, *The Kurzweil-Henstock Integral for Undergraduates*,
Compact Textbooks in Mathematics,
https://doi.org/10.1007/978-3-319-95321-2_1

Along this chapter, we denote by I a compact interval of the real line $\mathbb{R}$, i.e., an interval of the type $[a, b]$.

1.1 P-Partitions and Riemann Sums

Let us start by introducing the notion of P-partition of the interval I.

Definition 1.1
A **P-partition** of the interval $I = [a, b]$ is a set

$$\Pi = \{(x_1, [a_0, a_1]), (x_2, [a_1, a_2]), \dots, (x_m, [a_{m-1}, a_m])\},$$

whose elements appear as couples $(x_j, [a_{j-1}, a_j])$, where $[a_{j-1}, a_j]$ is a subset of I and x_j is a point in it. Precisely, we have

$$a = a_0 < a_1 < \cdots < a_{m-1} < a_m = b,$$

and, for every $j = 1, \dots, m$,

$$x_j \in [a_{j-1}, a_j].$$

Example Consider the interval $[0, 1]$. As examples of P-partitions of I we have the following sets:

$$\Pi = \left\{\left(\frac{1}{6}, [0, 1]\right)\right\}$$

$$\Pi = \left\{\left(0, \left[0, \frac{1}{3}\right]\right), \left(\frac{1}{2}, \left[\frac{1}{3}, 1\right]\right)\right\}$$

$$\Pi = \left\{\left(\frac{1}{3}, \left[0, \frac{1}{3}\right]\right), \left(\frac{1}{3}, \left[\frac{1}{3}, \frac{2}{3}\right]\right), \left(\frac{2}{3}, \left[\frac{2}{3}, 1\right]\right)\right\}$$

$$\Pi = \left\{\left(\frac{1}{8}, \left[0, \frac{1}{4}\right]\right), \left(\frac{3}{8}, \left[\frac{1}{4}, \frac{1}{2}\right]\right), \left(\frac{5}{8}, \left[\frac{1}{2}, \frac{3}{4}\right]\right), \left(\frac{7}{8}, \left[\frac{3}{4}, 1\right]\right)\right\}.$$

We consider now a function f, defined on the interval I, having real values. To each P-partition of the interval I we can associate a real number, in the following way.

Definition 1.2

Let $f : I \to \mathbb{R}$ be a function and

$$\Pi = \{(x_1, [a_0, a_1]), (x_2, [a_1, a_2]), \ldots, (x_m, [a_{m-1}, a_m])\}$$

a P-partition of I. We call **Riemann sum** associated to I, f and Π the real number $S(I, f, \Pi)$ defined by

$$S(I, f, \Pi) = \sum_{j=1}^{m} f(x_j)(a_j - a_{j-1}).$$

In order to better understand this definition, assume for simplicity the function f to be positive on I. Then, to each P-partition of I we associate the sum of the areas of the rectangles having base $[a_{j-1}, a_j]$ and height $[0, f(x_j)]$.

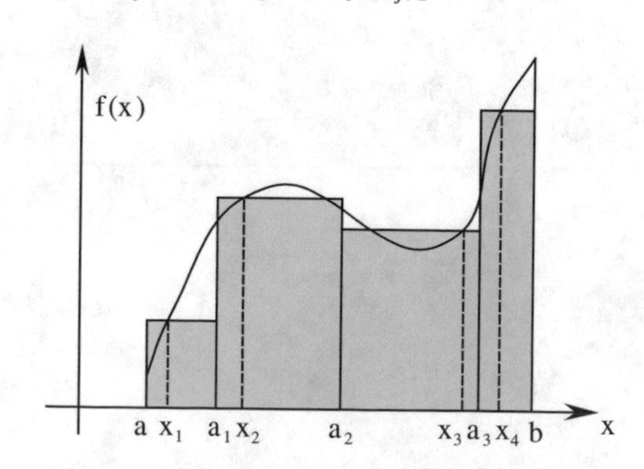

If f is not positive on I, the areas will be considered with positive or negative sign depending on whether $f(x_j)$ be positive or negative, respectively. If $f(x_j) = 0$, the j-th term of the sum will obviously be zero.

Example Let $I = [0, 1]$, $f(x) = 4x^2 - 1$, and

$$\Pi = \left\{ \left(\frac{1}{8}, \left[0, \frac{1}{4} \right] \right), \left(\frac{1}{2}, \left[\frac{1}{4}, \frac{3}{4} \right] \right), \left(\frac{7}{8}, \left[\frac{3}{4}, 1 \right] \right) \right\}.$$

Then,

$$S(I, f, \Pi) = -\frac{15}{16} \cdot \frac{1}{4} + 0 \cdot \frac{1}{2} + \frac{33}{16} \cdot \frac{1}{4} = \frac{9}{32}.$$

Now we ask whether, taking the P-partitions finer and finer, the Riemann sums associated to them will converge to some value. When this happens for a positive function f, such a value can be visualized as the area of the region in the Cartesian plane which is confined between the graph of f and the horizontal axis. To be able to analyze this question, we need to specify what we mean for a P-partition to be "fine".

1.2 The Notion of δ-Fineness

Let us introduce the notion of "fineness" for the P-partition Π previously defined. For brevity, we call **gauge** on I every function $\delta : I \to \mathbb{R}$ such that $\delta(x) > 0$ for every $x \in I$. Such a function will be useful for having a control on the amplitude of the various intervals determined by the points of the P-partition.

Definition 1.3

Given a gauge δ on I, we say that the P-partition Π introduced above is **δ-fine** if, for every $j = 1, \dots, m$,

$$x_j - a_{j-1} \le \delta(x_j), \quad \text{and} \quad a_j - x_j \le \delta(x_j).$$

Equivalently, we may write

$$[a_{j-1}, a_j] \subseteq [x_j - \delta(x_j), x_j + \delta(x_j)],$$

or else

$$x_j - \delta(x_j) \le a_{j-1} \le x_j \le a_j \le x_j + \delta(x_j).$$

We will show now that it is always possible to find a δ-fine P-partition of the interval I, whatever the gauge δ. In the following theorem, due to P. Cousin, the compactness of the interval I plays an essential role.

Theorem 1.4

Given a compact interval I, for every gauge δ on I there is a δ-fine P-partition of I.

Proof

We proceed by contradiction. Assume there exists a gauge δ on I for which it is impossible to find any δ-fine P-partition of I. Let us divide the interval I in two equal sub-intervals, having the mid point of I as common extremum. Then, at least one of the two sub-intervals does not have any δ-fine P-partition. Let us choose it, and divide it again in two equal sub-intervals. Continuing this way, we construct a sequence $(I_n)_n$ of bottled sub-intervals, whose lengths tend to zero, each of which does not have any δ-fine P-partitions. By the Cantor Theorem there is a point $c \in I$ belonging to all of these intervals. Moreover, it is clear that, from a certain n thereof, every I_n will be contained in $[c - \delta(c), c + \delta(c)]$. Choose one of these, e.g. $I_{\bar{n}}$. Then the set $\Pi = \{(c, I_{\bar{n}})\}$, whose only element is the couple $(c, I_{\bar{n}})$, is a δ-fine P-partition of $I_{\bar{n}}$, in contradiction with the above. □

Examples Let us see, as examples, some δ-fine P-partitions of the interval $[0, 1]$. We start with a constant gauge: $\delta(x) = \frac{1}{5}$. Since the previous theorem does not give any information on how to find a δ-fine P-partition, we will proceed by guessing. As a first guess, we choose the a_j equally distant and the x_j as the middle points of the intervals $[a_{j-1}, a_j]$. Hence:

$$a_j = \frac{j}{m}, \quad x_j = \frac{2j-1}{2m} \quad (j = 1, \dots, m).$$

For the corresponding P-partition to be δ-fine, it has to be

$$x_j - a_{j-1} = \frac{1}{2m} \leq \frac{1}{5}, \quad a_j - x_j = \frac{1}{2m} \leq \frac{1}{5}.$$

These inequalities are satisfied choosing $m \geq 3$. If $m = 3$, we have the δ-fine P-partition

$$\left\{\left(\frac{1}{6}, \left[0, \frac{1}{3}\right]\right), \left(\frac{1}{2}, \left[\frac{1}{3}, \frac{2}{3}\right]\right), \left(\frac{5}{6}, \left[\frac{2}{3}, 1\right]\right)\right\}.$$

If, instead of taking the points x_j in the middle of the respective intervals, we would like to choose them, for example, at the left extremum, i.e. $x_j = \frac{j-1}{m}$, in order to have a δ-fine P-partition we should ask that

$$x_j - a_{j-1} = 0 \leq \frac{1}{5}, \quad a_j - x_j = \frac{1}{m} \leq \frac{1}{5}.$$

These inequalities are verified if $m \geq 5$. For instance, if $m = 5$, we have the δ-fine P-partition

$$\left\{\left(0, \left[0, \frac{1}{5}\right]\right), \left(\frac{1}{5}, \left[\frac{1}{5}, \frac{2}{5}\right]\right), \left(\frac{2}{5}, \left[\frac{2}{5}, \frac{3}{5}\right]\right), \left(\frac{3}{5}, \left[\frac{3}{5}, \frac{4}{5}\right]\right), \left(\frac{4}{5}, \left[\frac{4}{5}, 1\right]\right)\right\}.$$

Notice that, with such a choice of the a_j, if $m \geq 5$, the points x_j can actually be taken arbitrarily in the respective intervals $[a_{j-1}, a_j]$, still obtaining δ-fine P-partitions.

The previous example shows how it is possible to construct δ-fine P-partitions in the case of a gauge δ which is constant with value $\frac{1}{5}$. It is clear that a similar procedure can be used for a constant gauge with arbitrary value. Consider now the case when δ is a continuous function. Then, the Weierstrass Theorem says that there is for $\delta(x)$ a minimum positive value: let it be $\bar{\delta}$. Consider then the constant gauge with value $\bar{\delta}$, and construct a $\bar{\delta}$-fine P-partition with the procedure we have seen above. Clearly, such a P-partition has to be δ-fine, as well. We have thus seen how the case of a continuous gauge can be reduced to that of a constant gauge.

Consider now the following non-continuous gauge:

$$\delta(x) = \begin{cases} \dfrac{1}{2} & \text{if } x = 0\,, \\ \dfrac{x}{2} & \text{if } x \in \,]0, 1]\,. \end{cases}$$

As before, we proceed by guessing. Let us try, as above, taking the a_j equally distant and the x_j as the middle points of the intervals $[a_{j-1}, a_j]$. This time, however, we are going to fail; indeed, we should have

$$x_1 = x_1 - a_0 \leq \delta(x_1) = \frac{x_1}{2}\,,$$

which is clearly impossible if $x_1 > 0$. The only way to solve this problem is to choose $x_1 = 0$. We decide then, for instance, to take the x_j to coincide with a_{j-1}, as was also done above. We thus find a δ-fine P-partition:

$$\left\{\left(0, \left[0, \frac{1}{2}\right]\right), \left(\frac{1}{2}, \left[\frac{1}{2}, \frac{3}{4}\right]\right), \left(\frac{3}{4}, \left[\frac{3}{4}, 1\right]\right)\right\}.$$

Notice that a more economic choice could have been

$$\left\{\left(0, \left[0, \frac{1}{2}\right]\right), \left(1, \left[\frac{1}{2}, 1\right]\right)\right\}.$$

The choice $x_1 = 0$ is however unavoidable.

Consider now the following gauge: once fixed a point $c \in \,]0, 1[$,

$$\delta(x) = \begin{cases} \dfrac{c - x}{2} & \text{if } x \in [0, c[\,, \\ \dfrac{1}{5} & \text{if } x = c\,, \\ \dfrac{x - c}{2} & \text{if } x \in \,]c, 1]\,. \end{cases}$$

Similar considerations to those made in the previous case lead to the conclusion that, in order to have a δ-fine P-partition, it is necessary that one of the x_j be chosen as to be the point c. For example, if $c = \frac{1}{2}$, a possible choice is the following:

$$\left\{\left(0,\left[0,\frac{1}{4}\right]\right),\left(\frac{1}{4},\left[\frac{1}{4},\frac{3}{8}\right]\right),\left(\frac{1}{2},\left[\frac{3}{8},\frac{5}{8}\right]\right),\left(\frac{3}{4},\left[\frac{5}{8},\frac{3}{4}\right]\right),\left(1,\left[\frac{3}{4},1\right]\right)\right\}.$$

1.3 Integrable Functions on a Compact Interval

Consider a function f, defined on the interval $I = [a, b]$. We are now in the position to define what we mean by convergence of the Riemann sum when the P-partitions become "finer and finer".[1]

Definition 1.5

A function $f : I \to \mathbb{R}$ is said to be **integrable** if there is a real number A with the following property: given $\varepsilon > 0$, it is possible to find a gauge δ on I such that, for every δ-fine P-partition Π of I, it is

$$|S(I, f, \Pi) - A| \le \varepsilon .$$

We will also say that f is **integrable on** I.

Let us prove that there is at most one $A \in \mathbb{R}$ which verifies the conditions of the definition. If there were a second one, say A', we would have that, for every $\varepsilon > 0$ there would be two gauges δ and δ' on I associated respectively to A and A' by the definition. Define the gauge

$$\delta''(x) = \min\{\delta(x), \delta'(x)\} .$$

Once a δ''-fine P-partition Π of I is chosen, we have that Π is both δ-fine and δ'-fine, hence

$$|A - A'| \le |A - S(I, f, \Pi)| + |S(I, f, \Pi) - A'| \le 2\varepsilon .$$

Since this holds for every $\varepsilon > 0$, it necessarily has to be $A = A'$.

If $f : I \to \mathbb{R}$ is an integrable function, the only element $A \in \mathbb{R}$ verifying the conditions of the definition is called the **integral** of f on I and is denoted by one of the following symbols:

$$\int_I f , \qquad \int_a^b f , \qquad \int_I f(x)\,dx , \qquad \int_a^b f(x)\,dx .$$

[1] The following definition is due to J. Kurzweil and R. Henstock: see Appendix E for a historical overview.

The presence of the letter x in the notation introduced here has no independent importance. It could be replaced by any other letter $t, u, \alpha, \dots$, or by any other symbol, unless already used with another meaning. For reasons to be explained later on, we set, moreover,

$$\int_b^a f = -\int_a^b f, \quad \text{and} \quad \int_a^a f = 0.$$

Examples Consider the function $f : [a, b] \to \mathbb{R}$, with $0 \le a < b$, defined by $f(x) = x^n$, where n is a natural number. In case $n = 0$, we have a constant function of value 1. In that case, the Riemann sums are all equal to $b - a$, and one easily verifies that the function is integrable and

$$\int_a^b 1 = b - a.$$

If $n = 1$, given a P-partition Π of $[a, b]$, the Riemann sum associated is

$$S(I, f, \Pi) = \sum_{j=1}^{m} x_j(a_j - a_{j-1}).$$

In order to find a candidate for the integral, let us consider a particular P-partition where the x_j are the middle points of the intervals $[a_{j-1}, a_j]$. In this particular case, we have

$$\sum_{j=1}^{m} x_j(a_j - a_{j-1}) = \sum_{j=1}^{m} \frac{a_{j-1} + a_j}{2}(a_j - a_{j-1}) = \frac{1}{2}\sum_{j=1}^{m}(a_j^2 - a_{j-1}^2) = \frac{1}{2}(b^2 - a^2).$$

We want to prove now that the function $f(x) = x$ is integrable on $[a, b]$ and that its integral is really $\frac{1}{2}(b^2 - a^2)$. Fix $\varepsilon > 0$. Taken any P-partition Π, we have:

$$\begin{aligned}
\left|S(I, f, \Pi) - \frac{1}{2}(b^2 - a^2)\right| &= \left|\sum_{j=1}^{m} x_j(a_j - a_{j-1}) - \sum_{j=1}^{m} \frac{a_{j-1} + a_j}{2}(a_j - a_{j-1})\right| \\
&\le \sum_{j=1}^{m} \left|x_j - \frac{a_{j-1} + a_j}{2}\right| (a_j - a_{j-1}) \\
&\le \sum_{j=1}^{m} \frac{a_j - a_{j-1}}{2}(a_j - a_{j-1}).
\end{aligned}$$

If we choose the gauge δ to be constant with value $\frac{\varepsilon}{b-a}$, then, for every δ-fine P-partition Π we have:

$$\left|S(I, f, \Pi) - \frac{1}{2}(b^2 - a^2)\right| \le \sum_{j=1}^{m} \frac{a_j - a_{j-1}}{2} 2\delta = \frac{\varepsilon}{b - a}\sum_{j=1}^{m}(a_j - a_{j-1}) = \varepsilon.$$

The condition of the definition is thus verified with this choice of the gauge, and we have proved that

$$\int_a^b x\,dx = \frac{1}{2}(b^2 - a^2)\,.$$

If $n = 2$, it is more difficult to find a candidate for the integral. It can be found by choosing a particular P-partition where the x_j are $[\frac{1}{3}(a_{j-1}^2 + a_{j-1}a_j + a_j^2)]^{1/2}$; indeed, in this case, the Riemann sum is given by

$$\sum_{j=1}^{m} \frac{a_{j-1}^2 + a_{j-1}a_j + a_j^2}{3}(a_j - a_{j-1}) = \frac{1}{3}\sum_{j=1}^{m}(a_j^3 - a_{j-1}^3) = \frac{1}{3}(b^3 - a^3)\,.$$

At this point, it is possible to proceed like in the case $n = 1$ to prove that the function $f(x) = x^2$ is integrable on $[a, b]$ and that its integral is $\frac{1}{3}(b^3 - a^3)$: once fixed $\varepsilon > 0$, choose the constant gauge $\delta = \frac{\varepsilon}{2(b^2 - a^2)}$ so that, for any δ-fine P-partition,

$$\begin{aligned}\left|S(I, f, \Pi) - \frac{1}{3}(b^3 - a^3)\right| &\le \sum_{j=1}^{m}\left|x_j^2 - \frac{a_{j-1}^2 + a_{j-1}a_j + a_j^2}{3}\right|(a_j - a_{j-1}) \\ &\le \sum_{j=1}^{m}(a_j^2 - a_{j-1}^2)2\delta \\ &= \frac{\varepsilon}{b^2 - a^2}\sum_{j=1}^{m}(a_j^2 - a_{j-1}^2) = \varepsilon\,.\end{aligned}$$

We have thus proved that

$$\int_a^b x^2\,dx = \frac{1}{3}(b^3 - a^3)\,.$$

In general, it is possible to prove in an analogous way that every function $f(x) = x^n$ is integrable on $[a, b]$, and

$$\int_a^b x^n\,dx = \frac{1}{n+1}(b^{n+1} - a^{n+1})\,.$$

1.4 Elementary Properties of the Integral

Let f, g be two real functions defined on $I = [a, b]$, and $\alpha \in \mathbb{R}$ be a constant. It is easy to verify that, for every P-partition Π of I,

$$S(I, f + g, \Pi) = S(I, f, \Pi) + S(I, g, \Pi)\,,$$

and

$$S(I, \alpha f, \Pi) = \alpha S(I, f, \Pi)\,.$$

These linearity properties are inherited by the integral, as will be proved in the following two propositions.

Proposition 1.6
If f and g are integrable on I, then $f + g$ is integrable on I and one has

$$\int_I (f+g) = \int_I f + \int_I g\,.$$

Proof
Set $A_1 = \int_I f$ and $A_2 = \int_I g$. Being $\varepsilon > 0$ fixed, there are two gauges δ_1 and δ_2 on I such that, for every P-partition Π of I, if Π is δ_1-fine, then

$$|S(I, f, \Pi) - A_1| \le \frac{\varepsilon}{2}\,,$$

while if Π is δ_2-fine, then

$$|S(I, g, \Pi) - A_2| \le \frac{\varepsilon}{2}\,.$$

Let us define the gauge δ on I as $\delta(x) = \min\{\delta_1(x), \delta_2(x)\}$. Let Π be a δ-fine P-partition of I. It is thus both δ_1-fine and δ_2-fine, and we have:

$$\begin{aligned} |S(I, f+g, \Pi) - (A_1 + A_2)| &= |S(I, f, \Pi) - A_1 + S(I, g, \Pi) - A_2| \\ &\le |S(I, f, \Pi) - A_1| + |S(I, g, \Pi) - A_2| \\ &\le \frac{\varepsilon}{2} + \frac{\varepsilon}{2} = \varepsilon\,. \end{aligned}$$

This completes the proof. □

Proposition 1.7
If f is integrable on I and $\alpha \in \mathbb{R}$, then αf is integrable on I and one has

$$\int_I (\alpha f) = \alpha \int_I f\,.$$

Proof

If $\alpha = 0$, the result is obvious. If $\alpha \neq 0$, set $A = \int_I f$ and fix $\varepsilon > 0$. There is a gauge δ on I such that

$$|S(I, f, \Pi) - A| \leq \frac{\varepsilon}{|\alpha|},$$

for every δ-fine P-partition Π of I. Then, for every δ-fine P-partition Π of I, we have

$$|S(I, \alpha f, \Pi) - \alpha A| = |\alpha S(I, f, \Pi) - \alpha A| \leq |\alpha| \frac{\varepsilon}{|\alpha|} = \varepsilon,$$

and the proof is thus completed. □

We have just proved that the set of integrable functions on $[a, b]$ is a vector space and that the integral is a linear function on it.

Example Every polynomial function is integrable on an interval $[a, b]$. If for instance $f(x) = 2x^2 - 3x + 7$, we have

$$\int_a^b f = 2\int_a^b x^2\,dx - 3\int_a^b x\,dx + 7\int_a^b 1\,dx = \frac{2}{3}(b^3 - a^3) - \frac{3}{2}(b^2 - a^2) + 7(b - a).$$

We now study the behavior of the integral with respect to the order relation in $\mathbb{R}$.

Proposition 1.8

If f is integrable on I and $f(x) \geq 0$ for every $x \in I$, then

$$\int_I f \geq 0.$$

Proof

Fix $\varepsilon > 0$. There is a gauge δ on I such that

$$\left| S(I, f, \Pi) - \int_I f \right| \leq \varepsilon,$$

for every δ-fine P-partition Π of I. Hence,

$$\int_I f \geq S(I, f, \Pi) - \varepsilon \geq -\varepsilon,$$

being clearly $S(I, f, \Pi) \geq 0$. Since this is true for every $\varepsilon > 0$, it has to be $\int_I f \geq 0$, thus proving the result. □

Corollary 1.9
If f and g are integrable on I and $f(x) \leq g(x)$ for every $x \in I$, then

$$\int_I f \leq \int_I g\,.$$

Proof
It is sufficient to apply the preceding proposition to the function $g - f$. □

Corollary 1.10
If f and $|f|$ are integrable on I, then

$$\left|\int_I f\right| \leq \int_I |f|\,.$$

Proof
Applying the preceding corollary to the inequalities

$$-|f| \leq f \leq |f|\,,$$

we have

$$-\int_I |f| \leq \int_I f \leq \int_I |f|\,,$$

whence the conclusion. □

1.5 The Fundamental Theorem

The following theorem constitutes a link between the differential and the integral calculus. It is called the **Fundamental Theorem of differential an integral calculus**. More briefly, we will call it the Fundamental Theorem.

Theorem 1.11
Let $F : [a, b] \to \mathbb{R}$ be a differentiable function, and let f be its derivative: $F'(x) = f(x)$ for every $x \in [a, b]$. Then, f is integrable on $[a, b]$, and

$$\int_a^b f = F(b) - F(a)\,.$$

Proof

Fix $\varepsilon > 0$. For any $x \in I = [a, b]$, since $F'(x) = f(x)$, there is a $\delta(x) > 0$ such that, for every $u \in I \cap [x - \delta(x), x + \delta(x)]$, one has

$$|F(u) - F(x) - f(x)(u - x)| \leq \frac{\varepsilon}{b-a}|u - x| .$$

We thus have a gauge δ on I. Consider now a δ-fine P-partition of I,

$$\Pi = \{(x_1, [a_0, a_1]), (x_2, [a_1, a_2]), \ldots, (x_m, [a_{m-1}, a_m])\} .$$

Since, for every $j = 1, \ldots, m$,

$$x_j - \delta(x_j) \leq a_{j-1} \leq x_j \leq a_j \leq x_j + \delta(x_j) ,$$

one has

$$\begin{aligned}
&|F(a_j) - F(a_{j-1}) - f(x_j)(a_j - a_{j-1})| = \\
&\quad = |F(a_j) - F(x_j) - f(x_j)(a_j - x_j) + [F(x_j) - F(a_{j-1}) + f(x_j)(a_{j-1} - x_j)]| \\
&\quad \leq |F(a_j) - F(x_j) - f(x_j)(a_j - x_j)| + |F(a_{j-1}) - F(x_j) - f(x_j)(a_{j-1} - x_j)| \\
&\quad \leq \frac{\varepsilon}{b-a}(|a_j - x_j| + |a_{j-1} - x_j|) \\
&\quad = \frac{\varepsilon}{b-a}(a_j - x_j + x_j - a_{j-1}) \\
&\quad = \frac{\varepsilon}{b-a}(a_j - a_{j-1}) .
\end{aligned}$$

We deduce that

$$\begin{aligned}
\left| F(b) - F(a) - \sum_{j=1}^{m} f(x_j)(a_j - a_{j-1}) \right| &= \\
&= \left| \sum_{j=1}^{m} [F(a_j) - F(a_{j-1})] - \sum_{j=1}^{m} f(x_j)(a_j - a_{j-1}) \right| \\
&= \left| \sum_{j=1}^{m} [F(a_j) - F(a_{j-1}) - f(x_j)(a_j - a_{j-1})] \right| \\
&\leq \sum_{j=1}^{m} \left| F(a_j) - F(a_{j-1}) - f(x_j)(a_j - a_{j-1}) \right| \\
&\leq \sum_{j=1}^{m} \frac{\varepsilon}{b-a}(a_j - a_{j-1}) = \varepsilon ,
\end{aligned}$$

and the theorem is proved. □

1.6 Primitivable Functions

We introduce the concept of *primitive* of a given function.

Definition 1.12
A function $f : I \to \mathbb{R}$ is said to be **primitivable** if there is a differentiable function $F : I \to \mathbb{R}$ such that $F'(x) = f(x)$ for every $x \in I$. Such a function F is called a **primitive** of f.

The Fundamental Theorem establishes that all primitivable functions defined on a compact interval $I = [a, b]$ are integrable, and that their integral is easily computable, once a primitive is known. It can be reformulated as follows.

Theorem 1.13
Let $f : [a, b] \to \mathbb{R}$ be primitivable and let F be a primitive. Then f is integrable on $[a, b]$, and

$$\int_a^b f = F(b) - F(a) .$$

It is sometimes useful to denote the difference $F(b) - F(a)$ with the symbols

$$[F]_a^b , \quad [F(x)]_{x=a}^{x=b} ,$$

or variants of these as, for instance, $[F(x)]_a^b$, when no ambiguities arise.

Example Consider the function $f(x) = x^n$. It is easy to see that $F(x) = \frac{1}{n+1}x^{n+1}$ is a primitive. The Fundamental Theorem tells us that

$$\int_a^b x^n \, dx = \left[\frac{1}{n+1}x^{n+1}\right]_a^b = \frac{1}{n+1}(b^{n+1} - a^{n+1}) ,$$

a result we already obtained in a direct way in the case $0 \leq a < b$.

The fact that the difference $F(b) - F(a)$ does not depend from the chosen primitive is explained by the following proposition.

Proposition 1.14
Let $f : I \to \mathbb{R}$ be a primitivable function, and let F be one of its primitives. Then, a function $G : I \to \mathbb{R}$ is a primitive of f if and only if $F - G$ is a constant function on I.

Proof

If $F - G$ is constant, then

$$G'(x) = (F + (G - F))'(x) = F'(x) + (G - F)'(x) = F'(x) = f(x),$$

for every $x \in I$, and hence G is a primitive of f. On the other hand, if G is a primitive of f, we have

$$(F - G)'(x) = F'(x) - G'(x) = f(x) - f(x) = 0,$$

for every $x \in I$. Consequently, $F - G$ is constant on I. □

Notice that, if $f : I \to \mathbb{R}$ is a primitivable function, it is also primitivable on every sub-interval of I. In particular, it is integrable on every interval $[a, x] \subseteq I$, and therefore it is possible to define a function

$$x \mapsto \int_a^x f,$$

which we call the **indefinite integral** of f. We denote this function by one of the following symbols:

$$\int_a^{\cdot} f, \qquad \int_a^{\cdot} f(t)\, dt$$

(notice that in this last notation it is convenient to use a different letter from x for the variable of f; for instance, we have chosen here the letter t). The Fundamental Theorem tells us that, if F is a primitive of f, then, for every $x \in [a, b]$,

$$\int_a^x f = F(x) - F(a).$$

This fact yields, taking into account Proposition 1.14, that the function $\int_a^{\cdot} f$ is itself a primitive of f. We thus have the following

Corollary 1.15

Let $f : [a, b] \to \mathbb{R}$ be a primitivable function. Then, the indefinite integral $\int_a^{\cdot} f$ is one of its primitives: it is a function defined on $[a, b]$, differentiable and, for every $x \in [a, b]$, we have

$$\left(\int_a^{\cdot} f \right)'(x) = f(x).$$

Notice that the choice of the point a in the definition of $\int_a^{\cdot} f$ is by no way necessary. One could take any point $\omega \in I$ and consider the function $\int_\omega^{\cdot} f$. The conventions made on the integral with exchanged extrema are such that the above stated theorem still holds. Indeed, if F is a primitive of f, even if $x < \omega$ we have

$$\int_{\omega}^{x} f = -\int_{x}^{\omega} f = -(F(\omega) - F(x)) = F(x) - F(\omega)\,,$$

so that $\int_{\omega}^{\cdot} f$ is still a primitive of f.

We will denote the set of all primitives of f with one of the following symbols:

$$\int f\,, \qquad \int f(x)\,dx\,.$$

Concerning the use of x, an observation analogous to the one made for the integral can be made here, as well: it can be replaced by any other letter or symbol, with the due precautions. When applying the theory to practical problems, however, if F denotes a primitive of f, instead of correctly writing

$$\int f = \{F + c : c \in \mathbb{R}\}\,,$$

it is common to find improper expressions of the type

$$\int f(x)\,dx = F(x) + c\,,$$

where $c \in \mathbb{R}$ stands for an arbitrary constant; we will adapt to this habit, too. Let us make a list of primitives of some elementary functions:

$$\int e^x\,dx = e^x + c\,,$$

$$\int \sin x\,dx = -\cos x + c\,,$$

$$\int \cos x\,dx = \sin x + c\,,$$

$$\int x^\alpha\,dx = \frac{x^{\alpha+1}}{\alpha+1} + c \qquad (\alpha \neq -1)\,,$$

$$\int \frac{1}{x}\,dx = \ln|x| + c\,,$$

$$\int \frac{1}{1+x^2}\,dx = \arctan x + c\,,$$

$$\int \frac{1}{\sqrt{1-x^2}}\,dx = \arcsin x + c\,.$$

Notice that the definition of primitivable function makes sense even in some cases where f is not necessarily defined on a compact interval, and indeed the formulas above are valid on the natural domains of the considered functions.

Example Using the Fundamental Theorem, we find:

$$\int_0^{\pi} \sin x\, dx = [-\cos x]_0^{\pi} = -\cos \pi + \cos 0 = 2\,.$$

Notice that the presence of the arbitrary constant c can sometimes lead to apparently different results. For example, it is readily verified that one also has

$$\int \frac{1}{\sqrt{1-x^2}}\, dx = -\arccos x + c\,.$$

This is explained by the fact that $\arcsin x = \frac{\pi}{2} - \arccos x$ for every $x \in [-1, 1]$, and one should not think that here c refers to the same constant as the one appearing in the last formula of the above list.

One should be careful with the notation introduced for the primitives, which looks similar to that for the integral, even if the two concepts are completely different. Their relation comes from the Fundamental Theorem: we have

$$\int_{\omega}^{\cdot} f \in \int f\,,$$

with any $\omega \in I$, and

$$\int_a^b f = \left[\int_{\omega}^{\cdot} f\right]_a^b.$$

One could be tempted to write

$$\int_a^b f = \left[\int f(x)\, dx\right]_a^b;$$

actually the left term is a real number, while the right term is something we have not even defined (it could be the set whose only element is $\int_a^b f$). In the applications, however, one often abuses of these notations.

From the known properties of derivatives, one can easily prove the following proposition.

Proposition 1.16
Let f and g be two functions, primitivable on the interval I, and $\alpha \in \mathbb{R}$ be arbitrary. Let F and G be two primitives of f and g, respectively. Then

(Continued)

Proposition 1.16 (continued)

1. *$f+g$ is primitivable on I and $F+G$ is one of its primitives; we will briefly write*[2]

$$\int (f+g) = \int f + \int g\,;$$

2. *αf is primitivable on I and αF is one of its primitives; we will briefly write*

$$\int (\alpha f) = \alpha \int f\,.$$

As a consequence of this proposition we have that the set of primitivable functions on I is a vector space.

We conclude this section exhibiting an interesting class of integrable functions which are not primitivable. Let the function $f : [a,b] \to \mathbb{R}$ be such that the set

$$E = \{x \in [a,b] : f(x) \neq 0\}$$

is finite or countable (for instance, a function which is zero everywhere except at a point, or the Dirichlet function, defined by $f(x) = 1$ if x is rational, and $f(x) = 0$ if x is irrational).

Let us prove that such a function is integrable, with $\int_a^b f = 0$. Assume for definiteness that E be infinite (the case when E is finite can be treated in an analogous way). Being countable, we can write $E = \{e_n : n \in \mathbb{N}\}$. Once $\varepsilon > 0$ has been fixed, we construct a gauge δ on $[a,b]$ in this way: if $x \notin E$, we set $\delta(x) = 1$; if instead for a certain n it is $x = e_n$, we set

$$\delta(e_n) = \frac{\varepsilon}{2^{n+3}|f(e_n)|}\,.$$

Let now $\Pi = \{(x_1,[a_0,a_1]),\dots,(x_m,[a_{m-1},a_m])\}$ be a δ-fine P-partition of $[a,b]$. By the way f is defined, the associated Riemann sum becomes

$$S([a,b],f,\Pi) = \sum_{\{1\le j\le m\,:\,x_j\in E\}} f(x_j)(a_j - a_{j-1})\,.$$

[2]Here and in the following we use in an intuitive way the algebraic operations involving sets. To be precise, the sum of two sets A and B is defined as

$$A + B = \{a+b : a \in A,\ b \in B\}\,.$$

Now, $[a_{j-1}, a_j] \subseteq [x_j - \delta(x_j), x_j + \delta(x_j)]$, so that $a_j - a_{j-1} \leq 2\delta(x_j)$, and if x_j is in E it is $x_j = e_n$, for some $n \in \mathbb{N}$. To any such e_n can however correspond one or two points x_j, so that we will have

$$\left| \sum_{\{1 \leq j \leq m : x_j \in E\}} f(x_j)(a_j - a_{j-1}) \right| \leq 2 \sum_{n=0}^{\infty} |f(e_n)| 2\delta(e_n) = 4 \sum_{n=0}^{\infty} \frac{\varepsilon}{2^{n+3}} = \varepsilon .$$

This shows that f is integrable on $[a, b]$ and that $\int_a^b f = 0$.

Let us see now that, if E is non-empty, then f is not primitivable on $[a, b]$. Indeed, if it were, its indefinite integral $\int_a^{\cdot} f$ should be one of its primitives. Proceeding as above, one sees that, for every $x \in [a, b]$, one has $\int_a^x f = 0$. Then, being the derivative of a constant function, f should be identically zero, which is false.

1.7 Primitivation by Parts and by Substitution

We introduce here two methods frequently used for determining the primitives of certain functions. The first is known as the method of primitivation **by parts**.

Proposition 1.17

Let $F, G : I \to \mathbb{R}$ be two differentiable functions, and let f, g be the respective derivatives. One has that fG is primitivable on I if and only if such is Fg, in which case a primitive of fG is obtained subtracting from FG a primitive of Fg; we will briefly write:

$$\int fG = FG - \int Fg .$$

Proof

Being F and G differentiable, such is FG, as well, and we have

$$(FG)' = fG + Fg .$$

Being $(FG)'$ primitivable on I with primitive FG, the conclusion follows from Proposition 1.16.

□

Example We would like to find a primitive of the function $h(x) = xe^x$. Define the following functions: $f(x) = e^x$, $G(x) = x$, and consequently $F(x) = e^x$, $g(x) = 1$. Applying the formula given by the above proposition, we have:

$$\int e^x x \, dx = e^x x - \int e^x \, dx = xe^x - e^x + c ,$$

where c stands, as usual, for an arbitrary constant.

As an immediate consequence of Proposition 1.17, we have the rule of **integration by parts**:

$$\int_a^b fG = F(b)G(b) - F(a)G(a) - \int_a^b Fg\,.$$

Examples Applying the formula directly to the function $h(x) = xe^x$ of the previous example, we obtain

$$\int_0^1 e^x x\,dx = e^1 \cdot 1 - e^0 \cdot 0 - \int_0^1 e^x\,dx = e - [e^x]_0^1 = e - (e^1 - e^0) = 1\,.$$

Notice that we could attain the same result using the Fundamental Theorem, having already found that a primitive of h is given by $H(x) = xe^x - e^x$:

$$\int_0^1 e^x x\,dx = H(1) - H(0) = (e - e) - (0 - 1) = 1\,.$$

Let us see some more examples. Let $h(x) = \sin^2 x$. With the obvious choice of the functions f and G, we find

$$\begin{aligned}\int \sin^2 x\,dx &= -\cos x \sin x + \int \cos^2 x\,dx \\ &= -\cos x \sin x + \int (1 - \sin^2 x)\,dx \\ &= x - \cos x \sin x - \int \sin^2 x\,dx\,,\end{aligned}$$

from which we get

$$\int \sin^2 x\,dx = \frac{1}{2}(x - \cos x \sin x) + c\,.$$

Consider now the case of the function $h(x) = \ln x$, with $x > 0$. In order to apply the formula of primitivation by parts, we choose the functions $f(x) = 1$, $G(x) = \ln x$. In this way, we find

$$\int \ln x\,dx = x \ln x - \int x\frac{1}{x}\,dx = x \ln x - \int 1\,dx = x \ln x - x + c\,.$$

The second method we want to study is known as the method of primitivation **by substitution**.

Proposition 1.18

Let $\varphi : I \to \mathbb{R}$ be a differentiable function and $f : \varphi(I) \to \mathbb{R}$ be a primitivable function on the interval $\varphi(I)$, with primitive F. Then, the function $(f \circ \varphi)\varphi'$ is primitivable on I, and one of its primitives is given by $F \circ \varphi$. We will briefly write:

$$\int (f \circ \varphi)\varphi' = \left(\int f\right) \circ \varphi .$$

Proof

The composite function $F \circ \varphi$ is differentiable on I and

$$(F \circ \varphi)' = (F' \circ \varphi)\varphi' = (f \circ \varphi)\varphi' .$$

It follows that $(f \circ \varphi)\varphi'$ is primitivable on I, with primitive $F \circ \varphi$. □

Example We look for a primitive of the function $h(x) = xe^{x^2}$. Defining $\varphi(x) = x^2$, $f(t) = \frac{1}{2}e^t$ (it is advisable to use different letters to indicate the variables of φ and f), we have that $h = (f \circ \varphi)\varphi'$. Since a primitive of f is seen to be $F(t) = \frac{1}{2}e^t$, a primitive of h is $F \circ \varphi$, i.e.

$$\int xe^{x^2}\, dx = F(\varphi(x)) + c = \frac{1}{2}e^{x^2} + c .$$

The formula of primitivation by substitution is often written in the form

$$\int f(\varphi(x))\varphi'(x)\, dx = \int f(t)\, dt \bigg|_{t=\varphi(x)} ,$$

where, if F is a primitive of f, the right term should be read as

$$\int f(t)\, dt \bigg|_{t=\varphi(x)} = F(\varphi(x)) + c ,$$

where $c \in \mathbb{R}$ is arbitrary. Formally, there is a "change of variable" $t = \varphi(x)$, and the symbol dt joins the game to replace $\varphi'(x)\, dx$ (the Leibniz notation $\frac{dt}{dx} = \varphi'(x)$ may be used as a mnemonic rule).

Example In order to find a primitive of the function $h(x) = \frac{\ln x}{x}$, we can choose $\varphi(x) = \ln x$, apply the formula

$$\int \frac{\ln x}{x}\, dx = \int t\, dt \bigg|_{t=\ln x} ,$$

and thus find $\frac{1}{2}(\ln x)^2 + c$ (in this case, writing $t = \ln x$, one has that the symbol dt replaces $\frac{1}{x}dx$).

As a consequence of the above formulas, we have the rule of **integration by substitution**:

$$\int_a^b f(\varphi(x))\varphi'(x)\,dx = \int_{\varphi(a)}^{\varphi(b)} f(t)\,dt\,.$$

Indeed, if F is a primitive of f on $\varphi(I)$, by the Fundamental Theorem, we have

$$\int_a^b (f \circ \varphi)\varphi' = (F \circ \varphi)(b) - (F \circ \varphi)(a) = F(\varphi(b)) - F(\varphi(a)) = \int_{\varphi(a)}^{\varphi(b)} f.$$

Example Taking the function $h(x) = xe^{x^2}$ considered above, and defining $\varphi(x) = x^2$, $f(t) = \frac{1}{2}e^t$, we have

$$\int_0^2 xe^{x^2}\,dx = \int_0^4 \frac{1}{2}e^t\,dt = \frac{1}{2}[e^t]_0^4 = \frac{e^4 - 1}{2}\,.$$

Clearly, the same result is obtainable directly by the Fundamental Theorem, once we know that a primitive of h is given by $H(x) = \frac{1}{2}e^{x^2}$. Indeed, we have

$$\int_0^2 xe^{x^2}\,dx = H(2) - H(0) = \frac{1}{2}e^4 - \frac{1}{2}e^0 = \frac{e^4 - 1}{2}\,.$$

In case the function $\varphi : I \to \varphi(I)$ be invertible, one can also write

$$\int f(t)\,dt = \int f(\varphi(x))\varphi'(x)\,dx\bigg|_{x=\varphi^{-1}(t)}\,,$$

with the corresponding formula for the integral,

$$\int_\alpha^\beta f(t)\,dt = \int_{\varphi^{-1}(\alpha)}^{\varphi^{-1}(\beta)} f(\varphi(x))\varphi'(x)\,dx\,.$$

Example Looking for a primitive of $f(t) = \sqrt{1 - t^2}$, with $t \in]-1, 1[$, we may try to consider the function $\varphi :]0, \pi[\to]-1, 1[$ defined as $\varphi(x) = \cos x$, so that

$$f(\varphi(x))\varphi'(x) = \sqrt{1 - \cos^2 x}\,(-\sin x) = -\sin^2 x\,.$$

As we have already proved, this last function is primitivable, so we can write

$$\begin{aligned}\int \sqrt{1-t^2}\,dt &= -\int \sin^2 x\,dx\Big|_{x=\arccos t}\\ &= -\frac{1}{2}(x-\sin x\cos x)\Big|_{x=\arccos t} + c\\ &= -\frac{1}{2}\left(\arccos t - t\sqrt{1-t^2}\right) + c\,.\end{aligned}$$

We are now in the position to compute primitives and integrals for a large class of functions. Some of these are proposed in the exercises below.

Exercises

1. Making use of the known rules for the computation of the primitives, recover the following formulas:

$$\int \frac{1}{(2+3x)^7}\,dx = -\frac{1}{18(2+3x)^6} + c\,,$$

$$\int \sqrt{x+7}\,dx = \frac{2}{3}\sqrt{(x+7)^3} + c\,,$$

$$\int \frac{x^2+3x-2}{\sqrt{x}}\,dx = \frac{2}{5}\sqrt{x^5} + 2\sqrt{x^3} - 4\sqrt{x} + c\,,$$

$$\int \frac{1}{\sqrt{x}-\sqrt{x+1}}\,dx = -\frac{2}{3}\left(\sqrt{(x+1)^3} + \sqrt{x^3}\right) + c\,,$$

$$\int \frac{1}{x^2-5x+6}\,dx = \ln|x-3| - \ln|x-2| + c\,,$$

$$\int \frac{1}{x^2+4x+5}\,dx = \arctan(x+2) + c\,,$$

$$\int \frac{1}{\sin^2 x\cos^2 x}\,dx = \tan x - \frac{1}{\tan x} + c\,,$$

$$\int \frac{1}{\cosh x}\,dx = 2\arctan(e^x) + c\,,$$

$$\int \frac{\ln x}{x}\,dx = \frac{1}{2}(\ln x)^2 + c\,.$$

2. Primitivation by parts gives the following:

$$\int x \sin x \, dx = \sin x - x \cos x + c\,,$$

$$\int \frac{\sqrt{1-x^2}}{x^2} \, dx = -\frac{\sqrt{1-x^2}}{x} + \arcsin x + c\,,$$

$$\int (\ln x)^2 \, dx = x\Big[(\ln x)^2 - 2\ln x + 2\Big] + c\,,$$

$$\int \arcsin x \, dx = x \arcsin x + \sqrt{1-x^2} + c\,.$$

3. Let $f : \mathbb{R} \to \mathbb{R}$ be a primitivable T-periodic function. Provide a criterion to ensure that its primitives are T-periodic, as well.
4. Prove that, if $f : \mathbb{R} \to \mathbb{R}$ is any primitivable function, then

$$\int_0^{2\pi} f(\sin x) \cos x \, dx = 0\,, \qquad \int_0^{2\pi} f(\cos x) \sin x \, dx = 0\,.$$

5. Given a primitivable function $f : \mathbb{R} \to \mathbb{R}$, prove that:
   a) if f is an odd function, then all its primitives are even functions;
   b) if f is an even function, then one of its primitives is an odd function;
   c) if $\int_a^b f = 0$ for every $a, b \in \mathbb{R}$, then f is identically equal to zero.

1.8 The Cauchy Criterion

We have the following characterization for a function to be integrable.

Theorem 1.19
A function $f : I \to \mathbb{R}$ is integrable if and only if for every $\varepsilon > 0$ there is a gauge δ on I such that, taking two δ-fine P-partitions Π, $\widetilde{\Pi}$ of I, one has

$$|S(I, f, \Pi) - S(I, f, \widetilde{\Pi})| \leq \varepsilon\,.$$

Proof
Let us see first the necessary condition. Let f be integrable on I with integral A, and fix $\varepsilon > 0$. Then, there is a gauge δ on I such that, for every δ-fine P-partition Π of I, it is

$$|S(I, f, \Pi) - A| \leq \frac{\varepsilon}{2}\,.$$

If Π and $\widetilde{\Pi}$ are two δ-fine P-partitions, we have:

$$|S(I, f, \Pi) - S(I, f, \widetilde{\Pi})| \leq |S(I, f, \Pi) - A| + |A - S(I, f, \widetilde{\Pi})| \leq \frac{\varepsilon}{2} + \frac{\varepsilon}{2} = \varepsilon .$$

Let us see now the sufficient condition. Once assumed the stated condition, let us choose $\varepsilon = 1$ so that we find a gauge δ_1 on I such that

$$|S(I, f, \Pi) - S(I, f, \widetilde{\Pi})| \leq 1 ,$$

whenever Π and $\widetilde{\Pi}$ are δ_1-fine P-partitions of I. Taking $\varepsilon = 1/2$, we can find a gauge δ_2 on I, that we can choose so that $\delta_2(x) \leq \delta_1(x)$ for every $x \in I$, such that

$$|S(I, f, \Pi) - S(I, f, \widetilde{\Pi})| \leq \frac{1}{2} ,$$

whenever Π and $\widetilde{\Pi}$ are δ_2-fine P-partitions of I. We can continue this way, choosing $\varepsilon = 1/k$, with k a positive integer, and find a sequence $(\delta_k)_k$ of gauges on I such that, for every $x \in I$,

$$\delta_1(x) \geq \delta_2(x) \geq \dots \geq \delta_k(x) \geq \delta_{k+1}(x) \geq \dots ,$$

and such that

$$|S(I, f, \Pi) - S(I, f, \widetilde{\Pi})| \leq \frac{1}{k} ,$$

whenever Π and $\widetilde{\Pi}$ are δ_k-fine P-partitions of I.

Let us fix, for every k, a δ_k-fine P-partition Π_k of I. We want to show that $(S(I, f, \Pi_k))_k$ is a Cauchy sequence of real numbers. Let $\bar{\varepsilon} > 0$ be given, and let us choose a positive integer m such that $m\bar{\varepsilon} \geq 1$. If $k_1 \geq m$ and $k_2 \geq m$, assuming for instance $k_2 \geq k_1$, we have

$$|S(I, f, \Pi_{k_1}) - S(I, f, \Pi_{k_2})| \leq \frac{1}{k_1} \leq \frac{1}{m} \leq \bar{\varepsilon} .$$

This proves that $(S(I, f, \Pi_k))_k$ is a Cauchy sequence. Hence, it has a finite limit, which we denote by A.

Now we show that A is just the integral of f on I. Fix $\varepsilon > 0$, let n be a positive integer such that $n\varepsilon \geq 1$, and consider the gauge $\delta = \delta_n$. For every δ-fine P-partition Π of I and for every $k \geq n$, it is

$$|S(I, f, \Pi) - S(I, f, \Pi_k)| \leq \frac{1}{n} \leq \varepsilon .$$

Letting k tend to $+\infty$, we have that $S(I, f, \Pi_k)$ tends to A, and consequently

$$|S(I, f, \Pi) - A| \leq \varepsilon .$$

The proof is thus completed. □

1.9 Integrability on Sub-Intervals

In this section we will see that if a function is integrable on an interval $I = [a, b]$, it is also integrable on any of its sub-intervals. In particular, it is possible to consider its indefinite integral. Moreover, we will see that, if a function is integrable on two contiguous intervals, it is also integrable on their union. More precisely, we have the following property of **additivity on sub-intervals**.

Theorem 1.20
Given three points $a < c < b$, let $f : [a, b] \to \mathbb{R}$ be a function. Then, f is integrable on $[a, b]$ if and only if it is integrable both on $[a, c]$ and on $[c, b]$. In this case,

$$\int_a^b f = \int_a^c f + \int_c^b f .$$

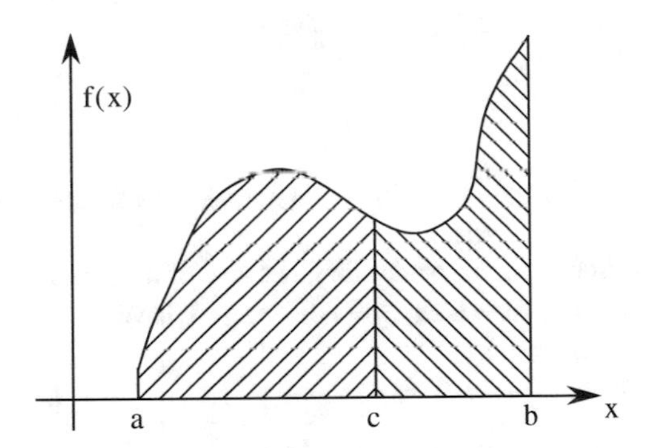

Proof
First assume f to be integrable on $[a, b]$. Let us choose for example the first sub-interval, $[a, c]$, and prove that f is integrable on it, using the Cauchy criterion. Fix $\varepsilon > 0$; being f integrable on $[a, b]$, it verifies the Cauchy condition on $[a, b]$, and hence there is a gauge δ on $[a, b]$ such that

$$|S([a, b], f, \Pi) - S([a, b], f, \widetilde{\Pi})| \leq \varepsilon ,$$

for every two δ-fine P-partitions Π, $\widetilde{\Pi}$ of $[a, b]$. The restrictions of δ to $[a, c]$ and $[c, b]$ are two gauges δ_1 and δ_2 on the respective sub-intervals. Let now Π_1 and $\widetilde{\Pi}_1$ be two δ_1-fine P-partitions of $[a, c]$. Let us fix a δ_2-fine P-partition Π_2 of $[c, b]$ and consider the P-partition Π of $[a, b]$ made by $\Pi_1 \cup \Pi_2$, and the P-partition $\widetilde{\Pi}$ of $[a, b]$ made by $\widetilde{\Pi}_1 \cup \Pi_2$. It is clear that both Π and $\widetilde{\Pi}$ are δ-fine. Moreover, we have

$$|S([a, c], f, \Pi_1) - S([a, c], f, \widetilde{\Pi}_1)| = |S([a, b], f, \Pi) - S([a, b], f, \widetilde{\Pi})| \leq \varepsilon ;$$

the Cauchy condition thus holds, so that f is integrable on $[a, c]$. Analogously it can be proved that f is integrable on $[c, b]$.

Suppose now that f be integrable on $[a, c]$ and on $[c, b]$, and let us prove then that f is integrable on $[a, b]$ with integral $\int_a^c f + \int_c^b f$. Once $\varepsilon > 0$ is fixed, there are a gauge δ_1 on $[a, c]$ and a gauge δ_2 on $[c, b]$ such that, for every δ_1-fine P-partition Π_1 of $[a, c]$ it is

$$\left| S([a, c], f, \Pi_1) - \int_a^c f \right| \leq \frac{\varepsilon}{2},$$

and for every δ_2-fine P-partition Π_2 of $[c, b]$ it is

$$\left| S([c, b], f, \Pi_2) - \int_c^b f \right| \leq \frac{\varepsilon}{2}.$$

We define now a gauge δ on $[a, b]$ in this way:

$$\delta(x) = \begin{cases} \min\left\{\delta_1(x), \frac{c-x}{2}\right\} & \text{if } a \leq x < c \\ \min\{\delta_1(c), \delta_2(c)\} & \text{if } x = c \\ \min\left\{\delta_2(x), \frac{x-c}{2}\right\} & \text{if } c < x \leq b\,. \end{cases}$$

Let now

$$\Pi = \{(x_1, [a_0, a_1]), (x_2, [a_1, a_2]), \dots, (x_m, [a_{m-1}, a_m])\}$$

be a δ-fine P-partition of $[a, b]$. Notice that, because of the particular choice of the gauge δ, there must be a certain $\bar{j}$ for which $x_{\bar{j}} = c$. Hence, we have

$$\begin{aligned} S([a, b], f, \Pi) = {} & f(x_1)(a_1 - a_0) + \dots + f(x_{\bar{j}-1})(a_{\bar{j}-1} - a_{\bar{j}-2}) + \\ & + f(c)(c - a_{\bar{j}-1}) + f(c)(a_{\bar{j}} - c) + \\ & + f(x_{\bar{j}+1})(a_{\bar{j}+1} - a_{\bar{j}}) + \dots + f(x_m)(a_m - a_{m-1})\,. \end{aligned}$$

Let us set

$$\Pi_1 = \{(x_1, [a_0, a_1]), (x_2, [a_1, a_2]), \dots, (x_{\bar{j}-1}, [a_{\bar{j}-2}, a_{\bar{j}-1}]), (c, [a_{\bar{j}-1}, c])\}$$

and

$$\Pi_2 = \{(c, [c, a_{\bar{j}}]), (x_{\bar{j}+1}, [a_{\bar{j}}, a_{\bar{j}+1}]), \dots, (x_m, [a_{m-1}, a_m])\}$$

(but in case $a_{\bar{j}-1}$ or $a_{\bar{j}}$ coincide with c we will have to take away an element). Then Π_1 is a δ_1-fine P-partition of $[a, c]$ and Π_2 is a δ_2-fine P-partition of $[c, b]$, and we have

$$S([a, b], f, \Pi) = S([a, c], f, \Pi_1) + S([c, b], f, \Pi_2)\,.$$

Consequently,

$$
\begin{aligned}
\left| S([a,b], f, \Pi) - \left(\int_a^c f + \int_c^b f \right) \right| &\leq \\
&\leq \left| S([a,c], f, \Pi_1) - \int_a^c f \right| + \left| S([c,b], f, \Pi_2) - \int_c^b f \right| \\
&\leq \frac{\varepsilon}{2} + \frac{\varepsilon}{2} = \varepsilon \,,
\end{aligned}
$$

which completes the proof. □

Example Consider the function $f : [0, 2] \to \mathbb{R}$ defined by

$$
f(x) = \begin{cases} 2 & \text{if } x \in [0, 1]\,, \\ 3 & \text{if } x \subset]1, 2]\,. \end{cases}
$$

Being f constant on $[0, 1]$, it is integrable there, and $\int_0^1 f = 2$. Moreover, on the interval $[1, 2]$ the function f differs from a constant only in one point: we have that $f(x) - 3$ is zero except for $x = 1$. For what have been proved somewhat before, the function $f - 3$ is integrable on $[1, 2]$ with zero integral and so, being $f = (f - 3) + 3$, even f is integrable and $\int_1^2 f = 3$. In conclusion,

$$
\int_0^2 f(x)\,dx = \int_0^1 f(x)\,dx + \int_1^2 f(x)\,dx = 2 + 3 = 5\,.
$$

It is easy to see from the theorem just proved above that if a function is integrable on an interval I, it still is on any sub-interval of I. Moreover, we have the following

Corollary 1.21
If $f : I \to \mathbb{R}$ is integrable, for any three arbitrarily chosen points u, v, w in I one has

$$
\int_u^w f = \int_u^v f + \int_v^w f\,.
$$

Proof
The case $u < v < w$ follows immediately from the previous theorem. The other possible cases are easily obtained using the conventions on the integrals with exchanged or equal extrema. □

1.10 R-Integrable Functions and Continuous Functions

Let us introduce an important class of integrable functions.

Definition 1.22

We say that an integrable function $f : I \to \mathbb{R}$ is **R-integrable** (or integrable according to Riemann), if among all possible gauges $\delta : I \to \mathbb{R}$ which verify the definition of integrability it is always possible to choose one which is constant on I.

It is immediate to see, repeating the proofs, that the set of R-integrable functions is a vector subspace of the space of integrable functions. Moreover, the following Cauchy criterion holds for R-integrable functions, whenever one considers only constant gauges.

Theorem 1.23

A function $f : I \to \mathbb{R}$ is R-integrable if and only if for every $\varepsilon > 0$ there is a constant $\delta > 0$ such that, taken two δ-fine P-partitions Π, $\widetilde{\Pi}$ of I, one has

$$|S(I, f, \Pi) - S(I, f, \widetilde{\Pi})| \leq \varepsilon .$$

The Cauchy criterion permits to prove the integrability of continuous functions. To simplify the expressions to come, we will denote by $\mu(K)$ the length of a bounded interval K. In particular,

$$\mu([a, b]) = b - a .$$

It will be useful, moreover, to make the convention that the length of the empty set is equal to zero.

Theorem 1.24

Every continuous function $f : I \to \mathbb{R}$ is R-integrable.

Proof

Fix $\varepsilon > 0$. Being f continuous on a compact interval, it is uniformly continuous there, so that there is a $\delta > 0$ such that, for x and x' in I,

$$|x - x'| \leq 2\delta \quad \Rightarrow \quad |f(x) - f(x')| \leq \frac{\varepsilon}{b - a} .$$

We will verify the Cauchy condition for the R-integrability by considering the constant gauge δ. Let

$$\Pi = \{(x_1, [a_0, a_1]), (x_2, [a_1, a_2]), \ldots, (x_m, [a_{m-1}, a_m])\}$$

and

$$\widetilde{\Pi} = \{(\tilde{x}_1, [\tilde{a}_0, \tilde{a}_1]), (\tilde{x}_2, [\tilde{a}_1, \tilde{a}_2]), \ldots, (\tilde{x}_{\widetilde{m}}, [\tilde{a}_{\widetilde{m}-1}, \tilde{a}_{\widetilde{m}}])\}$$

be two δ-fine P-partitions of I. Let us define the intervals (perhaps empty or reduced to a single point)

$$I_{j,k} = [a_{j-1}, a_j] \cap [\tilde{a}_{k-1}, \tilde{a}_k] .$$

Then, we have

$$a_j - a_{j-1} = \sum_{k=1}^{\widetilde{m}} \mu(I_{j,k}) , \qquad \tilde{a}_k - \tilde{a}_{k-1} = \sum_{j=1}^{m} \mu(I_{j,k}) ,$$

and, if $I_{j,k}$ is non-empty, $|x_j - \tilde{x}_k| \leq 2\delta$. Hence,

$$\begin{aligned} |S(I, f, \Pi) - S(I, f, \widetilde{\Pi})| &= \left| \sum_{j=1}^{m} \sum_{k=1}^{\widetilde{m}} f(x_j)\mu(I_{j,k}) - \sum_{k=1}^{\widetilde{m}} \sum_{j=1}^{m} f(\tilde{x}_k)\mu(I_{j,k}) \right| \\ &= \left| \sum_{j=1}^{m} \sum_{k=1}^{\widetilde{m}} [f(x_j) - f(\tilde{x}_k)]\, \mu(I_{j,k}) \right| \\ &\leq \sum_{j=1}^{m} \sum_{k=1}^{\widetilde{m}} |f(x_j) - f(\tilde{x}_k)|\, \mu(I_{j,k}) \\ &\leq \sum_{j=1}^{m} \sum_{k=1}^{\widetilde{m}} \frac{\varepsilon}{b-a}\, \mu(I_{j,k}) = \varepsilon . \end{aligned}$$

Therefore, the Cauchy condition holds true, and the proof is completed. □

Concerning the continuous functions, even the following holds.

Theorem 1.25
Every continuous function $f : [a, b] \to \mathbb{R}$ is primitivable.

Proof

Being continuous, f is integrable on every sub-interval of $[a, b]$, so that we can consider the function $\int_a^{\cdot} f$, indefinite integral of f. Let us prove that $\int_a^{\cdot} f$ is a primitive of f, i.e., that taken a point x_0 in $[a, b]$, the derivative of $\int_a^{\cdot} f$ in x_0 coincides with $f(x_0)$. We first consider the case when $x_0 \in]a, b[$. We want to prove that

$$\lim_{h\to 0} \frac{1}{h} \left(\int_a^{x_0+h} f - \int_a^{x_0} f \right) = f(x_0) .$$

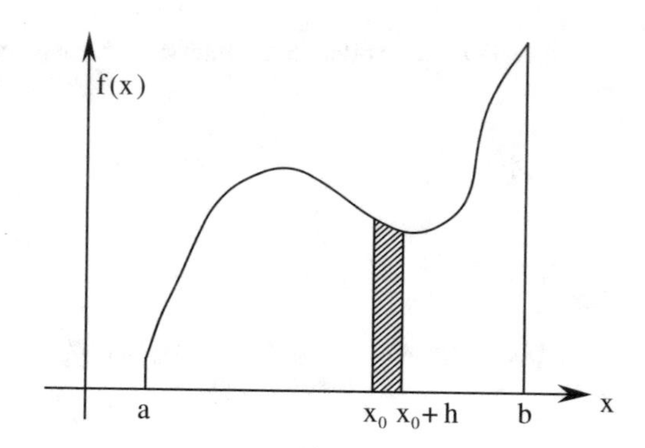

Equivalently, since

$$\frac{1}{h} \left(\int_a^{x_0+h} f - \int_a^{x_0} f \right) - f(x_0) = \frac{1}{h} \int_{x_0}^{x_0+h} (f(x) - f(x_0))\, dx ,$$

we will show that

$$\lim_{h\to 0} \frac{1}{h} \int_{x_0}^{x_0+h} (f(x) - f(x_0))\, dx = 0 .$$

Fix $\varepsilon > 0$. Being f continuous in x_0, there is a $\delta > 0$ such that, for every $x \in [a, b]$ satisfying $|x - x_0| \le \delta$, one has $|f(x) - f(x_0)| \le \varepsilon$. Taking h such that $0 < |h| \le \delta$, we distinguish two cases. If $0 < h \le \delta$, then

$$\left| \frac{1}{h} \int_{x_0}^{x_0+h} (f(x) - f(x_0))\, dx \right| \le \frac{1}{h} \int_{x_0}^{x_0+h} |f(x) - f(x_0)|\, dx \le \frac{1}{h} \int_{x_0}^{x_0+h} \varepsilon\, dx = \varepsilon ;$$

on the other hand, if $-\delta \le h < 0$, we have

$$\left| \frac{1}{h} \int_{x_0}^{x_0+h} (f(x) - f(x_0))\, dx \right| \le \frac{1}{-h} \int_{x_0+h}^{x_0} |f(x) - f(x_0)|\, dx \le \frac{1}{-h} \int_{x_0+h}^{x_0} \varepsilon\, dx = \varepsilon ,$$

and the proof is competed when $x_0 \in]a, b[$. In case $x_0 = a$ or $x_0 = b$, we proceed analogously, considering the right or the left derivative, respectively. □

Notice that it is not always possible to find an elementary expression for the primitive of a continuous function. As an example, the function $f(x) = e^{x^2}$ is primitivable, being continuous, but it can be proved that there is no elementary formula defining any of its primitives.[3]

Let us now prove that the Dirichlet function f is not R-integrable on any interval $[a, b]$ (remember that f is 1 on the rationals and 0 on the irrationals). We will show that the Cauchy condition is not verified. Take $\delta > 0$ and let $a = a_0 < a_1 < \cdots < a_m = b$ be such that, for every $j = 1, \ldots, m$, one has $a_j - a_{j-1} \leq \delta$. In every interval $[a_{j-1}, a_j]$ we can choose a rational number x_j and an irrational number $\tilde{x}_j$. The two P-partitions

$$\Pi = \{(x_1, [a_0, a_1]), (x_2, [a_1, a_2]), \ldots, (x_m, [a_{m-1}, a_m])\},$$

$$\widetilde{\Pi} = \{(\tilde{x}_1, [a_0, a_1]), (\tilde{x}_2, [a_1, a_2]), \ldots, (\tilde{x}_m, [a_{m-1}, a_m])\},$$

are δ-fine, and by the very definition of f we have

$$\begin{aligned} S([a,b], f, \Pi) - S([a,b], f, \widetilde{\Pi}) &= \sum_{j=1}^{m} [f(x_j) - f(\tilde{x}_j)](a_j - a_{j-1}) \\ &= \sum_{j=1}^{m} (a_j - a_{j-1}) = b - a\,. \end{aligned}$$

Since $\delta > 0$ has been taken arbitrarily, the Cauchy condition for R-integrability does not hold, so that f cannot be R-integrable on $[a, b]$.

Exercises

1. Recalling that

$$\sin(2\theta) = \frac{2\tan\theta}{1+\tan^2\theta}\,, \qquad \cos(2\theta) = \frac{1-\tan^2\theta}{1+\tan^2\theta}\,,$$

recover the following formulas:

$$\int \frac{1}{\sin x}\,dx = \ln\left|\tan\frac{x}{2}\right| + c\,, \qquad \int \frac{1}{\cos x}\,dx = \ln\left|\frac{1+\tan\frac{x}{2}}{1-\tan\frac{x}{2}}\right| + c\,.$$

Alternatively,

$$\int \frac{1}{\cos x}\,dx = \frac{1}{2}\ln\left|\frac{1+\sin x}{1-\sin x}\right| + c\,.$$

[3] By "elementary formula" I mean here an analytic formula where only polynomials, exponentials, logarithms and trigonometric functions appear.

2. Show that, if $f : [0, 1] \to \mathbb{R}$ is defined as

$$f(x) = \begin{cases} 1 & \text{if } x \notin \mathbb{Q}, \\ 3 & \text{if } x \in \mathbb{Q}, \end{cases}$$

then $\int_0^1 f(x)\,dx = 1$.
3. Let $f : \mathbb{R} \to \mathbb{R}$ be continuous and odd. Then, $\int_{-c}^{c} f = 0$, for every $c \in \mathbb{R}$.
4. Prove that, if $f : [a, b] \to \mathbb{R}$ is monotone, then it is R-integrable.
5. Show that every R-integrable function $f : [a, b] \to \mathbb{R}$ is necessarily bounded.

1.11 The Saks–Henstock Theorem

Let us go back to analyze the definition of integrability for a function $f : I \to \mathbb{R}$. The function f is integrable on I with integral $\int_I f$ if, for every $\varepsilon > 0$, there is a gauge δ on I such that, for every δ-fine P-partition

$$\Pi = \{(x_1, [a_0, a_1]), (x_2, [a_1, a_2]), \dots, (x_m, [a_{m-1}, a_m])\}$$

of I, the following holds:

$$\left| S(I, f, \Pi) - \int_I f \right| \le \varepsilon .$$

In this situation, since

$$S(I, f, \Pi) = \sum_{j=1}^{m} f(x_j)(a_j - a_{j-1}) , \qquad \int_I f = \sum_{j=1}^{m} \int_{a_{j-1}}^{a_j} f ,$$

we have

$$\left| \sum_{j=1}^{m} \left(f(x_j)(a_j - a_{j-1}) - \int_{a_{j-1}}^{a_j} f \right) \right| \le \varepsilon .$$

This fact tells us that the sum of all “errors” $(f(x_j)(a_j - a_{j-1}) - \int_{a_{j-1}}^{a_j} f)$ is arbitrarily small, provided that the P-partition be sufficiently fine. Notice that those “errors” may be either positive or negative, so that in the sum they could compensate with one another. The following **Saks–Henstock’s theorem** tells us that even the sum of all absolute values of those “errors” can be made arbitrarily small.

Theorem 1.26
Let $f : I \to \mathbb{R}$ be an integrable function and let δ be a gauge on I such that, for every δ-fine P-partition Π of I, it happens that $|S(I, f, \Pi) - \int_I f| \le \varepsilon$. Then, for such P-partitions $\Pi = \{(x_1, [a_0, a_1]), \ldots, (x_m, [a_{m-1}, a_m])\}$ we also have

$$\sum_{j=1}^{m} \left| f(x_j)(a_j - a_{j-1}) - \int_{a_{j-1}}^{a_j} f \right| \le 4\varepsilon .$$

Proof
We consider separately in the sum the positive and the negative terms. Let us prove that the sum of the positive terms is less than or equal to 2ε. In an analogous way one can proceed for the negative terms. Rearranging the terms in the sum, we can assume that the positive ones be the first q terms $(f(x_j)(a_j - a_{j-1}) - \int_{a_{j-1}}^{a_j} f)$, with $j = 1, \ldots, q$, i.e.,

$$f(x_1)(a_1 - a_0) - \int_{a_0}^{a_1} f \,, \;\ldots,\; f(x_q)(a_q - a_{q-1}) - \int_{a_{q-1}}^{a_q} f \,.$$

Consider the remaining $m - q$ intervals $[a_{k-1}, a_k]$, with $k = q + 1, \ldots, m$:

$$[a_q, a_{q+1}] \,, \;\ldots,\; [a_{m-1}, a_m] \,.$$

Being f integrable on these intervals, there exist some gauges δ_k on $[a_{k-1}, a_k]$, respectively, which we can choose such that $\delta_k(x) \le \delta(x)$ for every $x \in [a_{k-1}, a_k]$, for which

$$\left| S([a_{k-1}, a_k], f, \Pi_k) - \int_{a_{k-1}}^{a_k} f \right| \le \frac{\varepsilon}{m - q} \,,$$

for every δ_k-fine P-partition Π_k of $[a_{k-1}, a_k]$. Consequently, the family $\tilde{\Pi}$ made by the couples $(x_1, [a_0, a_1]), \ldots, (x_q, [a_{q-1}, a_q])$ and by the elements of the families Π_k, with k varying from $q + 1$ to m, is a δ-fine P-partition of I such that

$$S(I, f, \tilde{\Pi}) = \sum_{j=1}^{q} f(x_j)(a_j - a_{j-1}) + \sum_{k=q+1}^{m} S([a_{k-1}, a_k], f, \Pi_k) \,.$$

Then, we have:

$$\begin{aligned} \sum_{j=1}^{q} \left(f(x_j)(a_j - a_{j-1}) - \int_{a_{j-1}}^{a_j} f \right) &= \sum_{j=1}^{q} f(x_j)(a_j - a_{j-1}) - \sum_{j=1}^{q} \int_{a_{j-1}}^{a_j} f \\ &= \left(S(I, f, \tilde{\Pi}) - \sum_{k=q+1}^{m} S([a_{k-1}, a_k], f, \Pi_k) \right) - \left(\int_I f - \sum_{k=q+1}^{m} \int_{a_{k-1}}^{a_k} f \right) \end{aligned}$$

$$\leq \left| S(I, f, \widetilde{\Pi}) - \int_I f \right| + \sum_{k=q+1}^{m} \left| S([a_{k-1}, a_k], f, \Pi_k) - \int_{a_{k-1}}^{a_k} f \right|$$

$$\leq \varepsilon + (m-q)\frac{\varepsilon}{m-q} = 2\varepsilon .$$

Proceeding similarly for the negative terms, the conclusion follows. □

The following corollary will be useful in the next section to study the integrability of the absolute value of an integrable function.

Corollary 1.27
Let $f : I \to \mathbb{R}$ be an integrable function and let δ be a gauge on I such that, for every δ-fine P-partition Π of I, it happens that $|S(I, f, \Pi) - \int_I f| \leq \varepsilon$. Then, for such P-partitions $\Pi = \{(x_1, [a_0, a_1]), \dots, (x_m, [a_{m-1}, a_m])\}$ we also have

$$\left| S(I, |f|, \Pi) - \sum_{j=1}^{m} \left| \int_{a_{j-1}}^{a_j} f \right| \right| \leq 4\varepsilon .$$

Proof
Using the well known inequalities for the absolute value, we have:

$$\left| S(I, |f|, \Pi) - \sum_{j=1}^{m} \left| \int_{a_{j-1}}^{a_j} f \right| \right| = \left| \sum_{j=1}^{m} \left[|f(x_j)|(a_j - a_{j-1}) - \left| \int_{a_{j-1}}^{a_j} f \right| \right] \right|$$

$$\leq \sum_{j=1}^{m} \left| |f(x_j)(a_j - a_{j-1})| - \left| \int_{a_{j-1}}^{a_j} f \right| \right|$$

$$\leq \sum_{j=1}^{m} \left| f(x_j)(a_j - a_{j-1}) - \int_{a_{j-1}}^{a_j} f \right| \leq 4\varepsilon .$$

This completes the proof. □

In the sequel it will be useful to consider even the so-called **sub-P-partitions** of the interval I. A sub-P-partition is a set of the type $\Xi = \{(\xi_j, [\alpha_j, \beta_j]) : j = 1, \dots, m\}$, where the intervals $[\alpha_j, \beta_j]$ are non-overlapping, but not necessarily contiguous, and $\xi_j \in [\alpha_j, \beta_j]$ for every $j = 1, \dots, m$. Using the Cousin's theorem, it is easy to see that every sub-P-partition can be extended to a P-partition of I.

For a sub-P-partition Ξ, it is still meaningful to consider the associated Riemann sum:

$$S(I, f, \Xi) = \sum_{j=1}^{m} f(\xi_j)(\beta_j - \alpha_j).$$

Moreover, given a gauge δ on I, the sub-P-partition Ξ is δ-fine if, for every j, one has

$$\xi_j - \alpha_j \leq \delta(\xi_j) \quad \text{e} \quad \beta_j - \xi_j \leq \delta(\xi_j).$$

The Saks–Henstock's theorem can then be generalized to the sub-P-partitions, simply considering the fact that every sub-P-partition is a subset of a P-partition. We can thus obtain the following **equivalent statement of the Saks–Henstock's theorem**.

Theorem 1.28
Let $f : I \to \mathbb{R}$ be an integrable function and let δ be a gauge on I such that, for every δ-fine P-partition Π of I, it happens that $|S(I, f, \Pi) - \int_I f| \leq \varepsilon$. Then, for every δ-fine sub-P-partition $\Xi = \{(\xi_j, [\alpha_j, \beta_j]) : j = 1, \ldots, m\}$ of I, we have

$$\sum_{j=1}^{m} \left| f(\xi_j)(\beta_j - \alpha_j) - \int_{\alpha_j}^{\beta_j} f \right| \leq 4\varepsilon.$$

Notice that, as a consequence of this last statement, for such sub-P-partitions Ξ we have, in particular,

$$\left| S(I, f, \Xi) - \sum_{j=1}^{m} \int_{\alpha_j}^{\beta_j} f \right| \leq 4\varepsilon.$$

1.12 L-Integrable Functions

In this section, we introduce another important class of integrable functions on the interval $I = [a, b]$.

Definition 1.29
We say that an integrable function $f : I \to \mathbb{R}$ is **L-integrable** (or integrable according to Lebesgue), if even $|f|$ happens to be integrable on I.

It is clear that every positive integrable function is L-integrable. Moreover, every continuous function on $[a, b]$ is L-integrable there, since $|f|$ is still continuous. We have the following characterization of L-integrability.

Proposition 1.30
Let $f : I \to \mathbb{R}$ be an integrable function, and consider the set $\mathcal{S}$ of all real numbers

$$\sum_{i=1}^{q} \left| \int_{c_{i-1}}^{c_i} f \right| ,$$

obtained choosing $c_0, c_1, \dots, c_q$ in I in such a way that $a = c_0 < c_1 < \cdots < c_q = b$. The function f is L-integrable on I if and only if $\mathcal{S}$ is bounded above. In that case, we have

$$\int_I |f| = \sup \mathcal{S} .$$

Proof
Assume first f to be L-integrable on I. If $a = c_0 < c_1 < \cdots < c_q = b$, then f and $|f|$ are integrable on every sub-interval $[c_{i-1}, c_i]$, and we have

$$\sum_{i=1}^{q} \left| \int_{c_{i-1}}^{c_i} f \right| \leq \sum_{i=1}^{q} \int_{c_{i-1}}^{c_i} |f| = \int_I |f| .$$

Consequently, the set $\mathcal{S}$ is bounded above: $\sup \mathcal{S} \leq \int_I |f|$.

On the other hand, assume now $\mathcal{S}$ to be bounded above and let us prove that in that case $|f|$ is integrable on I and $\int_I |f| = \sup \mathcal{S}$. Fix $\varepsilon > 0$. Let δ_1 be a gauge such that, for every δ_1-fine P-partition Π of I, one has

$$\left| S(I, f, \Pi) - \int_I f \right| \leq \frac{\varepsilon}{8} .$$

Letting $A = \sup \mathcal{S}$, by the properties of the sup there surely are $a = c_0 < c_1 < \cdots < c_q = b$ such that

$$A - \frac{\varepsilon}{2} \leq \sum_{i=1}^{q} \left| \int_{c_{i-1}}^{c_i} f \right| \leq A .$$

We construct the gauge δ_2 in such a way that, for every $x \in I$, it has to be that $[x - \delta_2(x), x + \delta_2(x)]$ meets only those intervals $[c_{i-1}, c_i]$ to which x belongs. In this way,

- if x belongs to the interior of one of the intervals $[c_{i-1}, c_i]$, we have that $[x - \delta_2(x), x + \delta_2(x)]$ is contained in $]c_{i-1}, c_i[$;
- if x coincides with one of the c_i in the interior of $[a, b]$, then $[x - \delta_2(x), x + \delta_2(x)]$ is contained in $]c_{i-1}, c_{i+1}[$;
- if $x = a$, then $[x, x + \delta_2(x)]$ is contained in $[a, c_1[$;
- if $x = b$, then $[x - \delta_2(x), x]$ is contained in $]c_{q-1}, b]$.

Set, for every $x \in I, \delta(x) = \min\{\delta_1(x), \delta_2(x)\}$. Once taken a δ-fine P-partition $\Pi = \{(x_1, [a_0, a_1]), \dots, (x_m, [a_{m-1}, a_m])\}$ of I, consider the intervals (possibly empty or reduced to a point)

$$I_{j,i} = [a_{j-1}, a_j] \cap [c_{i-1}, c_i].$$

The choice of the gauge δ_2 yields that, if $I_{j,i}$ has a positive length, then $x_j \in I_{j,i}$. Indeed, if $x_j \notin [c_{i-1}, c_i]$, then

$$[a_{j-1}, a_j] \cap [c_{i-1}, c_i] \subseteq [x_j - \delta_2(x_j), x_j + \delta_2(x_j)] \cap [c_{i-1}, c_i] = \emptyset.$$

Therefore, taking those $I_{j,i}$, the set

$$\tilde{\Pi} = \{(x_j, I_{j,i}) : j = 1, \dots, m \ , \ i = 1, \dots, q \ , \ \mu(I_{j,i}) > 0\}$$

is a δ-fine P-partition of I, and we have

$$S(I, |f|, \Pi) = \sum_{j=1}^{m} |f(x_j)|(a_j - a_{j-1}) = \sum_{j=1}^{m} \sum_{i=1}^{q} |f(x_j)| \mu(I_{j,i}) = S(I, |f|, \tilde{\Pi}).$$

Moreover,

$$A - \frac{\varepsilon}{2} \le \sum_{i=1}^{q} \left| \int_{c_{i-1}}^{c_i} f \right| = \sum_{i=1}^{q} \left| \sum_{j=1}^{m} \int_{I_{j,i}} f \right| \le \sum_{i=1}^{q} \sum_{j=1}^{m} \left| \int_{I_{j,i}} f \right| \le A,$$

and by Corollary 1.27,

$$\left| S(I, |f|, \tilde{\Pi}) - \sum_{i=1}^{q} \sum_{j=1}^{m} \left| \int_{I_{j,i}} f \right| \right| \le 4 \frac{\varepsilon}{8} = \frac{\varepsilon}{2}.$$

Consequently, we have

$$\begin{aligned} |S(I, |f|, \Pi) - A| &= |S(I, |f|, \tilde{\Pi}) - A| \\ &\le \left| S(I, |f|, \tilde{\Pi}) - \sum_{i=1}^{q} \sum_{j=1}^{m} \left| \int_{I_{j,i}} f \right| \right| + \left| \sum_{i=1}^{q} \sum_{j=1}^{m} \left| \int_{I_{j,i}} f \right| - A \right| \\ &\le \frac{\varepsilon}{2} + \frac{\varepsilon}{2} = \varepsilon, \end{aligned}$$

which is what was to be proved. □

We have a series of corollaries.

Corollary 1.31
Let $f, g : I \to \mathbb{R}$ be two integrable functions such that, for every $x \in I$,

$$|f(x)| \leq g(x);$$

then f is L-integrable on I.

Proof
Take $c_0, c_1, \ldots, c_q$ in I so that $a = c_0 < c_1 < \cdots < c_q = b$. Being $-g(x) \leq f(x) \leq g(x)$ for every $x \in I$, we have that

$$-\int_{c_{i-1}}^{c_i} g \leq \int_{c_{i-1}}^{c_i} f \leq \int_{c_{i-1}}^{c_i} g\,,$$

i.e.

$$\left|\int_{c_{i-1}}^{c_i} f\right| \leq \int_{c_{i-1}}^{c_i} g\,,$$

for every $1 \leq i \leq q$. Hence,

$$\sum_{i=1}^{q}\left|\int_{c_{i-1}}^{c_i} f\right| \leq \sum_{i=1}^{q}\int_{c_{i-1}}^{c_i} g = \int_I g\,.$$

Then, the set $\mathcal{S}$ is bounded above by $\int_I g$, so that f is L-integrable on I. □

Corollary 1.32
Let $f, g : I \to \mathbb{R}$ be two L-integrable functions and $\alpha \in \mathbb{R}$ be a constant. Then $f + g$ and αf are L-integrable on I.

Proof
By the assumption, f, $|f|$ and g, $|g|$ are integrable on I. Then, such are also $f+g$, $|f|+|g|$, αf, and $|\alpha||f|$. On the other hand, for every $x \in I$, it is

$$|(f+g)(x)| \leq |f(x)| + |g(x)|\,,$$

$$|\alpha f(x)| \leq |\alpha|\,|f(x)|\,.$$

Corollary 1.31 then guarantees that $f + g$ and αf are L-integrable on I. □

We have thus proved that the L-integrable functions make up a vector subspace of the space of integrable functions.

Corollary 1.33
Let $f_1, f_2 : I \to \mathbb{R}$ be two L-integrable functions. Then $\min\{f_1, f_2\}$ *and* $\max\{f_1, f_2\}$ *are L-integrable on I.*

Proof
It follows immediately from the formulas

$$\min\{f_1, f_2\} = \frac{1}{2}(f_1 + f_2 - |f_1 - f_2|)\,,$$

$$\max\{f_1, f_2\} = \frac{1}{2}(f_1 + f_2 + |f_1 - f_2|)$$

and from Corollary 1.32. □

Corollary 1.34
A function $f : I \to \mathbb{R}$ is L-integrable if and only if both its positive part $f^+ =$ $\max\{f, 0\}$ *and its negative part $f^- = \max\{-f, 0\}$ are integrable on I. In that case,* $\int_I f = \int_I f^+ - \int_I f^-$.

Proof
It follows immediately from Corollary 1.33 and the formulas $f = f^+ - f^-$, $|f| = f^+ + f^-$. □

We want to see now an example of an integrable function which is not L-integrable. Let $f : [0, 1] \to \mathbb{R}$ be defined by

$$f(x) = \frac{1}{x} \sin \frac{1}{x^2}\,,$$

if $x \neq 0$, and $f(0) = 0$. Let us define the two auxiliary functions $g : [0, 1] \to \mathbb{R}$ and $h : [0, 1] \to \mathbb{R}$ such that, if $x \neq 0$,

$$g(x) = \frac{1}{x} \sin \frac{1}{x^2} + x \cos \frac{1}{x^2}\,, \qquad h(x) = -x \cos \frac{1}{x^2}\,,$$

and $g(0) = h(0) = 0$. It is easily seen that g is primitivable on $[0, 1]$ and that one of its primitives $G : [0, 1] \to \mathbb{R}$ is given by

$$G(x) = \frac{x^2}{2} \cos \frac{1}{x^2}\,,$$

if $x \neq 0$, and $G(0) = 0$. Moreover, h is continuous on $[0, 1]$, hence it is primitivable there, too. Hence, even the function $f = g + h$ is primitivable on $[0, 1]$. By the Fundamental Theorem, f is then integrable on $[0, 1]$. We will show now that $|f|$ is not integrable on $[0, 1]$. Consider the intervals $[((k+1)\pi)^{-1/2}, (k\pi)^{-1/2}]$, with $k \geq 1$. The function $|f|$ is continuous on these intervals, hence it is primitivable there. By the substitution $y = 1/x^2$, we obtain

$$\int_{((k+1)\pi)^{-1/2}}^{(k\pi)^{-1/2}} \frac{1}{x}\left|\sin\frac{1}{x^2}\right| dx = \int_{k\pi}^{(k+1)\pi} \frac{1}{2y}|\sin y|\,dy\,.$$

On the other hand,

$$\int_{k\pi}^{(k+1)\pi} \frac{1}{2y}|\sin y|\,dy \geq \frac{1}{2(k+1)\pi}\int_{k\pi}^{(k+1)\pi} |\sin y|\,dy = \frac{1}{(k+1)\pi}\,.$$

If $|f|$ were integrable on $[0, 1]$, we would have that, for every $n \geq 1$,

$$\begin{aligned}\int_0^1 |f| &= \int_0^{((n+1)\pi)^{-1/2}} |f| + \sum_{k=1}^{n} \int_{((k+1)\pi)^{-1/2}}^{(k\pi)^{-1/2}} |f| + \int_{\pi^{-1/2}}^{1} |f| \\ &\geq \sum_{k=1}^{n} \int_{((k+1)\pi)^{-1/2}}^{(k\pi)^{-1/2}} |f| \\ &\geq \sum_{k=1}^{n} \frac{1}{(k+1)\pi}\,,\end{aligned}$$

which is impossible, since the series $\sum_{k=1}^{\infty} \frac{1}{(k+1)\pi}$ diverges. Hence, f is not L-integrable on $[0, 1]$.

1.13 The Monotone Convergence Theorem

In this section and in the next one, we will consider the situation where a sequence of integrable functions $(f_k)_k$ converges pointwise to a function f : for every $x \in I$,

$$\lim_{k\to\infty} f_k(x) = f(x)\,.$$

The question is whether

$$\int_I f = \lim_{k\to\infty} \int_I f_k\,,$$

i.e., whether the following formula holds:

$$\int_I \Big(\lim_{k\to\infty} f_k(x)\Big)\,dx = \lim_{k\to\infty} \int_I f_k(x)\,dx\,.$$

Example Let us first show that in some cases the answer could be in the negative. Consider the functions $f_k : [0,\pi] \to \mathbb{R}$, with $k = 1, 2, \dots$, defined by

$$f_k(x) = \begin{cases} k\sin(kx) & \text{if } x \in [0, \frac{\pi}{k}]\,,\\ 0 & \text{otherwise.}\end{cases}$$

For every $x \in [0,\pi]$, we have $\lim_{k\to\infty} f_k(x) = 0$, while

$$\int_0^{\pi} f_k(x)\,dx = \int_0^{\pi/k} k\sin(kx)\,dx = \int_0^{\pi} \sin(t)\,dt = 2\,.$$

We will see now that the formula holds true if the sequence of functions is monotone, or bounded in some way. Let us start with the following result, known as the **Monotone Convergence Theorem**, due to B. Levi.

Theorem 1.35
We are given a function $f : I \to \mathbb{R}$ and a sequence of functions $f_k : I \to \mathbb{R}$, with $k \in \mathbb{N}$, verifying the following conditions:
1. *the sequence $(f_k)_k$ converges pointwise to f;*
2. *the sequence $(f_k)_k$ is monotone;*
3. *each function f_k is integrable on I;*
4. *the real sequence $(\int_I f_k)_k$ has a finite limit.*

Then, f is integrable on I, and

$$\int_I f = \lim_{k\to\infty} \int_I f_k\,.$$

Proof
We assume for definiteness the sequence $(f_k)_k$ to be increasing on I; therefore, we have

$$f_k(x) \le f_{k+1}(x) \le f(x)\,,$$

for every $k \in \mathbb{N}$ and every $x \in I$. Let us set

$$A = \lim_{k\to\infty} \int_I f_k \,.$$

We will prove that f is integrable on I and that A is its integral. Fix $\varepsilon > 0$. Being every f_k integrable on I, there are some gauges δ_k^* on I such that, if Π_k is a δ_k^*-fine P-partition of I,

$$\left| S(I, f_k, \Pi_k) - \int_I f_k \right| \leq \frac{\varepsilon}{3 \cdot 2^{k+3}} \,.$$

Moreover, there is a $\bar{k} \in \mathbb{N}$ such that, for every $k \geq \bar{k}$, it is

$$0 \leq A - \int_I f_k \leq \frac{\varepsilon}{3} \,,$$

and since the sequence $(f_k)_k$ converges pointwise on I to f, for every $x \in I$ there is a natural number $n(x) \geq \bar{k}$, such that, for every $k \geq n(x)$, one has

$$|f_k(x) - f(x)| \leq \frac{\varepsilon}{3(b-a)} \,.$$

Let us define the gauge δ in the following way: for every $x \in I$,

$$\delta(x) = \delta^*_{n(x)}(x) \,.$$

Let now $\Pi = \{(x_1, [a_0, a_1]), \dots, (x_m, [a_{m-1}, a_m])\}$ be a δ-fine P-partition of I. We have:

$$\begin{aligned}
|S(I, f, \Pi) \; - A| &= \left| \sum_{j=1}^m f(x_j)(a_j - a_{j-1}) - A \right| \\
&\leq \left| \sum_{j=1}^m [f(x_j) - f_{n(x_j)}(x_j)](a_j - a_{j-1}) \right| + \\
&\quad + \left| \sum_{j=1}^m \left[f_{n(x_j)}(x_j)(a_j - a_{j-1}) - \int_{a_{j-1}}^{a_j} f_{n(x_j)} \right] \right| + \left| \sum_{j=1}^m \int_{a_{j-1}}^{a_j} f_{n(x_j)} - A \right| .
\end{aligned}$$

Estimation of the first term gives

$$\begin{aligned}
\left| \sum_{j=1}^m [f(x_j) - f_{n(x_j)}(x_j)](a_j - a_{j-1}) \right| &\leq \sum_{j=1}^m |f(x_j) - f_{n(x_j)}(x_j)|(a_j - a_{j-1}) \\
&\leq \sum_{j=1}^m \frac{\varepsilon}{3(b-a)}(a_j - a_{j-1}) = \frac{\varepsilon}{3} \,.
\end{aligned}$$

In order to estimate the second term, set

$$r = \min_{1 \le j \le m} n(x_j) , \quad s = \max_{1 \le j \le m} n(x_j) ,$$

and notice that, putting together the terms whose indices $n(x_j)$ coincide with a same value k, by the second statement of Saks–Henstock's theorem (Theorem 1.28) we obtain

$$\begin{aligned}
&\left| \sum_{j=1}^{m} \left[f_{n(x_j)}(x_j)(a_j - a_{j-1}) - \int_{a_{j-1}}^{a_j} f_{n(x_j)} \right] \right| = \\
&\quad = \left| \sum_{k=r}^{s} \left\{ \sum_{\{1 \le j \le m \,:\, n(x_j) = k\}} \left[f_k(x_j)(a_j - a_{j-1}) - \int_{a_{j-1}}^{a_j} f_k \right] \right\} \right| \\
&\quad \le \sum_{k=r}^{s} \sum_{\{1 \le j \le m \,:\, n(x_j) = k\}} \left| f_k(x_j)(a_j - a_{j-1}) - \int_{a_{j-1}}^{a_j} f_k \right| \\
&\quad \le \sum_{k=r}^{s} 4 \frac{\varepsilon}{3 \cdot 2^{k+3}} \le \frac{\varepsilon}{3} .
\end{aligned}$$

Concerning the third term, since $r \ge \bar{k}$, using the monotonicity of the sequence $(f_k)_k$ we have

$$\begin{aligned}
0 &\le A - \int_I f_s = A - \sum_{j=1}^{m} \int_{a_{j-1}}^{a_j} f_s \le \\
&\le A - \sum_{j=1}^{m} \int_{a_{j-1}}^{a_j} f_{n(x_j)} \le \\
&\le A - \sum_{j=1}^{m} \int_{a_{j-1}}^{a_j} f_r = A - \int_I f_r \le \frac{\varepsilon}{3} ,
\end{aligned}$$

from which

$$\left| \sum_{j=1}^{m} \int_{a_{j-1}}^{a_j} f_{n(x_j)} - A \right| \le \frac{\varepsilon}{3} .$$

Hence,

$$|S(I, f, \Pi) - A| \le \frac{\varepsilon}{3} + \frac{\varepsilon}{3} + \frac{\varepsilon}{3} = \varepsilon ,$$

and the proof is thus completed. □

As an immediate consequence of the Monotone Convergence Theorem, we have the analogous statement for a series of functions.

Corollary 1.36
We are given a function $f : I \to \mathbb{R}$ *and a sequence of functions* $f_k : I \to \mathbb{R}$, *with* $k \in \mathbb{N}$, *verifying the following conditions:*

1. *the series* $\sum_k f_k$ *converges pointwise to* f;
2. *for every* $k \in \mathbb{N}$ *and every* $x \in I$, *it is* $f_k(x) \geq 0$;
3. *each function* f_k *is integrable on* I;
4. *the series* $\sum_k (\int_I f_k)$ *converges.*
5. *Then,* f *is integrable on* I, *and*

$$\int_I f = \sum_{k=0}^{\infty} \int_I f_k \,.$$

Example Consider the Taylor series associated to the function $f(x) = e^{x^2}$,

$$e^{x^2} = \sum_{k=0}^{\infty} \frac{x^{2k}}{k!} \,.$$

The functions $f_k(x) = \frac{x^{2k}}{k!}$ satisfy the assumptions 1 and 2 of the corollary. Moreover, they are integrable on $I = [a, b]$ and

$$\int_a^b f_k(x)\,dx = \left[\frac{x^{2k+1}}{(2k+1)k!} \right]_a^b = \frac{b^{2k+1} - a^{2k+1}}{(2k+1)k!} \,,$$

so that it can be seen that the series $\sum_k (\int_I f_k)$ converges. It is then possible to apply the corollary, thus obtaining

$$\int_a^b e^{x^2}\,dx = \sum_{k=0}^{\infty} \frac{b^{2k+1} - a^{2k+1}}{(2k+1)k!} \,.$$

In particular, considering the indefinite integral $\int_0^{\cdot} f$, we find an expression for the primitives of e^{x^2}, i.e.,

$$\int e^{x^2}\,dx = \sum_{k=0}^{\infty} \frac{x^{2k+1}}{(2k+1)k!} + c \,.$$

1.14 The Dominated Convergence Theorem

We start by proving the following preliminary result.

Lemma 1.37
Let $f_1, f_2, \dots, f_n : I \to \mathbb{R}$ be integrable functions. If there exists an integrable function $g : I \to \mathbb{R}$ such that, for every $x \in I$ and $1 \le k \le n$ it happens that

$$g(x) \le f_k(x),$$

then $\min\{f_1, f_2, \dots, f_n\}$ *and* $\max\{f_1, f_2, \dots, f_n\}$ *are integrable on I.*

Proof
Consider the case $n = 2$. The functions $f_1 - g$ and $f_2 - g$, being integrable and non-negative, are L-integrable. Hence, $\min\{f_1 - g, f_2 - g\}$ and $\max\{f_1 - g, f_2 - g\}$ are L-integrable, by Corollary 1.33. The conclusion then follows from the fact that

$$\min\{f_1, f_2\} = \min\{f_1 - g, f_2 - g\} + g,$$

$$\max\{f_1, f_2\} = \max\{f_1 - g, f_2 - g\} + g.$$

The general case can be easily obtained by induction. □

We are now ready to state and prove the following important result due to H. Lebesgue, known as the **Dominated Convergence Theorem**.

Theorem 1.38
We are given a function $f : I \to \mathbb{R}$ and a sequence of functions $f_k : I \to \mathbb{R}$, with $k \in \mathbb{N}$, verifying the following conditions:

1. *the sequence $(f_k)_k$ converges pointwise to f;*
2. *each function f_k is integrable on I;*
3. *there are two integrable functions $g, h : I \to \mathbb{R}$ for which*

$$g(x) \le f_k(x) \le h(x),$$

for every $k \in \mathbb{N}$ and $x \in I$.

Then, the sequence $(\int_I f_k)_k$ has a finite limit, f is integrable on I, and

$$\int_I f = \lim_{k \to \infty} \int_I f_k.$$

Proof

For any couple of natural numbers k, ℓ, define the functions

$$\phi_{k,\ell} = \min\{f_k, f_{k+1}, \dots, f_{k+\ell}\}, \quad \Phi_{k,\ell} = \max\{f_k, f_{k+1}, \dots, f_{k+\ell}\}.$$

By the above proved lemma, all $\phi_{k,\ell}$ and $\Phi_{k,\ell}$ are integrable on I. Moreover, for any fixed k, the sequence $(\phi_{k,\ell})_\ell$ is decreasing and bounded below by g, and the sequence $(\Phi_{k,\ell})_\ell$ is increasing and bounded above by h. hence, these sequences converge to two functions ϕ_k and Φ_k, respectively:

$$\lim_{\ell\to\infty} \phi_{k,\ell} = \phi_k = \inf\{f_k, f_{k+1}, \dots\},$$

$$\lim_{\ell\to\infty} \Phi_{k,\ell} = \Phi_k = \sup\{f_k, f_{k+1}, \dots\}.$$

Furthermore, the sequence $(\int_I \phi_{k,\ell})_\ell$ is decreasing and bounded below by $\int_I g$, while the sequence $(\int_I \Phi_{k,\ell})_\ell$ is increasing and bounded above by $\int_I h$. The Monotone Convergence Theorem then guarantees that the functions ϕ_k and Φ_k are integrable on I.

Now, the sequence $(\phi_k)_k$ is increasing, and the sequence $(\Phi_k)_k$ is decreasing; as $\lim_{k\to\infty} f_k = f$, we must have

$$\lim_{k\to\infty} \phi_k = \liminf_{k\to\infty} f_k = f,$$

$$\lim_{k\to\infty} \Phi_k = \limsup_{k\to\infty} f_k = f.$$

Moreover, the sequence $(\int_I \phi_k)_k$ is increasing and bounded above by $\int_I h$, while the sequence $(\int_I \Phi_k)_k$ is decreasing and bounded below by $\int_I g$. We can then apply again the Monotone Convergence Theorem, from which we deduce that f is integrable on I and

$$\int_I f = \lim_{k\to\infty} \int_I \phi_k = \lim_{k\to\infty} \int_I \Phi_k.$$

Being $\phi_k \le f_k \le \Phi_k$, we have $\int_I \phi_k \le \int_I f_k \le \int_I \Phi_k$, and the conclusion follows. □

Example Consider, for $k \ge 1$, the functions $f_k : [0,3] \to \mathbb{R}$ defined by $f_k(x) = \arctan\left(kx - \frac{k^2}{k+1}\right)$. We have the following situation:

$$\lim_{k\to\infty} f_k(x) = \begin{cases} -\dfrac{\pi}{2} & \text{if } x \in [0,1[\,, \\ 0 & \text{if } x = 1\,, \\ \dfrac{\pi}{2} & \text{if } x \in \,]1,3]\,. \end{cases}$$

Moreover,

$$|f_k(x)| \le \frac{\pi}{2},$$

for every $k \in \mathbb{N}$ and $x \in [0, 3]$. The assumptions of the theorem are then satisfied, taking the two constant functions $g(x) = -\frac{\pi}{2}$, $h(x) = \frac{\pi}{2}$. We can then conclude that

$$\lim_{k\to\infty} \int_0^3 \arctan\left(kx - \frac{k^2}{k+1}\right) dx = -\frac{\pi}{2} + 2\frac{\pi}{2} = \frac{\pi}{2}.$$

Exercises

1. By the use of the Dominated Convergence Theorem, prove that

$$\lim_{k\to\infty} \int_0^1 \frac{\sin(e^{kx})}{\sqrt{k}}\, dx = 0, \qquad \lim_{k\to\infty} \int_{-1}^1 \sqrt{k} \arctan\left(\frac{x}{k}\right) dx = 0.$$

2. Let $f_k : [0, 1] \to \mathbb{R}$ be defined as

$$f_k(x) = \begin{cases} 0 & \text{if } x \in \left[0, \frac{1}{k}\right[, \\ \frac{1}{kx} & \text{if } x \in \left[\frac{1}{k}, 1\right]. \end{cases}$$

Prove that

$$\lim_{k\to\infty} \int_0^1 f_k(x)\, dx = 0,$$

both by a direct computation and by the Monotone or the Dominated Convergence Theorems. Let now

$$f_k(x) = \begin{cases} 0 & \text{if } x \in \left[0, \frac{1}{k^2}\right[, \\ \frac{1}{kx} & \text{if } x \in \left[\frac{1}{k^2}, 1\right]. \end{cases}$$

Explain why, in this case, neither the Monotone nor the Dominated Convergence Theorems can be applied.

3. Compute the following limit:

$$\lim_{k\to\infty} \int_0^\pi k^3 \left(\sin\left(\frac{x}{k}\right) - \frac{x}{k}\right) dx.$$

1.15 Integration on Non-Compact Intervals

We begin by considering a function $f : [a, b[\to \mathbb{R}$, where $b \leq +\infty$. Assume that f be integrable on every compact interval of the type $[a, c]$, with $c \in]a, b[$. This happens, for instance, when f is continuous on $[a, b[$.

Definition 1.39

We say that a function $f : [a, b[\to \mathbb{R}$ is **integrable** if f is integrable on $[a, c]$ for every $c \in]a, b[$, and the limit

$$\lim_{c \to b^-} \int_a^c f$$

exists and is finite. In that case, the above limit is called the **integral** of f on $[a, b[$ and it is denoted by $\int_a^b f$, or $\int_a^b f(x)\, dx$.

In particular, if $b = +\infty$, we will write: $\int_a^{+\infty} f$, or $\int_a^{+\infty} f(x)\, dx$.

Examples Let $a > 0$; it is readily seen that the function $f : [a, +\infty[\to \mathbb{R}$, defined by $f(x) = x^{-\alpha}$, is integrable if and only if $\alpha > 1$, in which case we have

$$\int_a^{+\infty} \frac{dx}{x^\alpha} = \frac{a^{1-\alpha}}{\alpha - 1} .$$

Consider now the case $a < b < +\infty$. It can be verified that the function $f : [a, b[\to \mathbb{R}$, defined by $f(x) = (b - x)^{-\beta}$, is integrable if and only if $\beta < 1$, in which case we have

$$\int_a^b \frac{dx}{(b - x)^\beta} = \frac{(b - a)^{1-\beta}}{1 - \beta} .$$

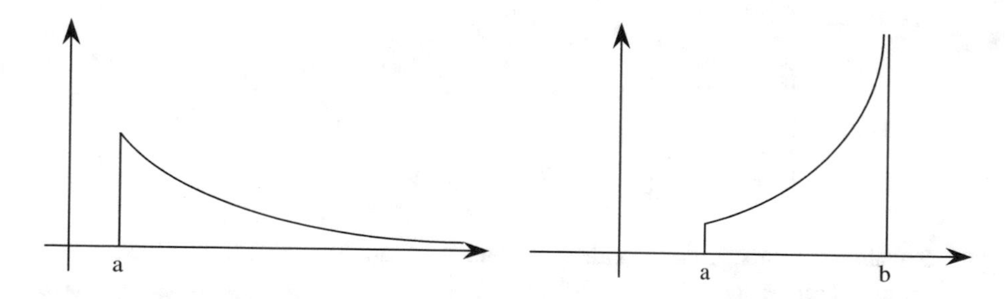

One often speaks of **improper integral** when considering functions which are defined on non-compact intervals. One also says that *the integral converges* if the function f is integrable on $[a, b[$, i.e., when the limit $\lim_{c \to b^-} \int_a^c f$ exists and is finite. If the limit does not exist, it is said that *the integral is undetermined*. If it exists and equals $+\infty$ or $-\infty$, it is said that *the integral diverges* to $+\infty$ or to $-\infty$, respectively.

It is clear that the convergence of the integral depends solely on the behavior of the function near the point b. In other words, modifying the function outside a neighborhood of b, the convergence of the integral is by no means compromised.

Let us now state the **Cauchy convergence criterion**.

Theorem 1.40
Let $f : [a, b[\to \mathbb{R}$ be a function, which is integrable on $[a, c]$, for every $c \in]a, b[$. A necessary and sufficient condition for f to be integrable on $[a, b[$ is that for every $\varepsilon > 0$ there is a $\bar{c} \in]a, b[$ such that, taking as c' and c'' any two numbers in $[\bar{c}, b[$, it is

$$\left| \int_{c'}^{c''} f \right| \leq \varepsilon .$$

Proof
It is a direct consequence of the Cauchy criterion for the limit, when applied to the function $F : [a, b[\to \mathbb{R}$ defined as $F(c) = \int_a^c f$. □

From the Cauchy criterion we deduce the following **comparison criterion**.

Theorem 1.41
Let $f : [a, b[\to \mathbb{R}$ be a function, which is integrable on $[a, c]$, for every $c \in]a, b[$. If there is an integrable function $g : [a, b[\to \mathbb{R}$ such that, for every $x \in [a, b[$,

$$|f(x)| \leq g(x) ,$$

then f is integrable on $[a, b[$, too.

Proof
Once fixed $\varepsilon > 0$, there is a $\bar{c} \in]a, b[$ such that, taking arbitrarily c', c'' in $[\bar{c}, b[$, it is $|\int_{c'}^{c''} g| \leq \varepsilon$. If for example $c' \leq c''$, being $-g \leq f \leq g$, one has

$$-\int_{c'}^{c''} g \leq \int_{c'}^{c''} f \leq \int_{c'}^{c''} g ,$$

and therefore

$$\left| \int_{c'}^{c''} f \right| \leq \int_{c'}^{c''} g \leq \varepsilon .$$

The Cauchy condition then holds, whence the conclusion. □

As an immediate consequence, we have the following.

Corollary 1.42
Let $f : [a, b[\to \mathbb{R}$ be a function, which is integrable on $[a, c]$, for every $c \in]a, b[$. If $|f|$ is integrable on $[a, b[$, then also f is such, and

$$\left| \int_a^b f \right| \leq \int_a^b |f| .$$

In the case when both f and $|f|$ are integrable on $[a, b[$, we say that f is L-integrable, or absolutely integrable, on $[a, b[$.

Example Consider the function $f : [\pi, +\infty[\to \mathbb{R}$ defined by $f(x) = \frac{\sin x}{x}$. We will see that it is integrable on $[\pi, +\infty[$, but not absolutely integrable there. To see that it is integrable, take $c > \pi$ and integrate by parts:

$$\int_\pi^c \frac{\sin x}{x}\, dx = \left[\frac{-\cos x}{x} \right]_\pi^c - \int_\pi^c \frac{\cos x}{x^2}\, dx .$$

We find that

$$\lim_{c \to +\infty} \int_\pi^c f = -\frac{1}{\pi} - \lim_{c \to +\infty} \int_\pi^c \frac{\cos x}{x^2}\, dx ,$$

and this last limit is finite by the comparison theorem, since

$$\left| \frac{\cos x}{x^2} \right| \leq x^{-2} .$$

Hence, f is integrable on $[\pi, +\infty[$. Assume by contradiction that it was also absolutely integrable. In that case, for every integer $n \geq 2$, we would have

$$\begin{aligned} \int_\pi^{n\pi} \left| \frac{\sin x}{x} \right| dx &= \sum_{k=1}^{n-1} \int_{k\pi}^{(k+1)\pi} \frac{|\sin x|}{x}\, dx \\ &\geq \sum_{k=1}^{n-1} \frac{1}{(k+1)\pi} \int_{k\pi}^{(k+1)\pi} |\sin x|\, dx \\ &= \frac{2}{\pi} \sum_{k=1}^{n-1} \frac{1}{k+1} . \end{aligned}$$

but this is impossible, since the series $\sum_{k=1}^\infty \frac{1}{k+1}$ diverges.

Let us now state a corollary of the comparison criterion which is often used in practice.

Corollary 1.43
Let $f, g : [a, b[\to \mathbb{R}$ be two functions with positive values, which are integrable on $[a, c]$ for every $c \in]a, b[$. Assume that the following limit exists:

$$L = \lim_{x \to b^-} \frac{f(x)}{g(x)}.$$

Then, the following conclusions hold:
a) if $L = 0$ and g is integrable on $[a, b[$, then such is f, as well;
b) if $0 < L < +\infty$, then f is integrable on $[a, b[$ if and only if such is g;
c) if $L = +\infty$ and g is not integrable on $[a, b[$, then neither f is such.

Example Consider the function $f : [0, +\infty[\to \mathbb{R}$ defined by

$$f(x) = e^{1/(x^2+1)} - 1.$$

As a comparison function, I would like to take $g(x) = x^{-2}$. A technical problem arises, however, since g is not defined on $[0, +\infty[$. We can proceed in two different ways: either we restrict f to an interval of the type $[a, +\infty[$, with $a > 0$, and we observe that this operation does not modify the convergence (or the non convergence) of the integral, since f is continuous on $[0, a]$; or we adapt to this situation the function g : for instance, we can choose

$$g(x) = \begin{cases} 1 & \text{if } x \in [0, 1], \\ x^{-2} & \text{if } x \geq 1. \end{cases}$$

Once this has been done, observe that

$$\lim_{x \to +\infty} \frac{f(x)}{g(x)} = \lim_{x \to +\infty} x^2 (e^{1/(x^2+1)} - 1) = \lim_{x \to +\infty} \frac{x^2}{x^2+1} \frac{e^{1/(x^2+1)} - 1}{1/(x^2+1)} = 1.$$

Since g is integrable on $[0, +\infty[$, such is f, as well.

We consider now a function $f :]a, b] \to \mathbb{R}$, with $a \geq -\infty$. There is an analogous definition of its improper integral.

Definition 1.44
We say that a function $f :]a, b] \to \mathbb{R}$ is **integrable** if f is integrable on $[c, b]$ for every $c \in]a, b[$, and the limit

$$\lim_{c \to a^+} \int_c^b f$$

(Continued)

Definition 1.44 (continued)

exists and is finite. In that case, the above limit is called the **integral** of f on $]a, b]$ and it is denoted by $\int_a^b f$, or $\int_a^b f(x)\,dx$.

Given the function $f :]a, b] \to \mathbb{R}$, it is possible to consider the function $g : [a', b'[\to \mathbb{R}$, with $a' = -b$ and $b' = -a$, defined by $g(x) = f(-x)$. It is easy to see that f is integrable on $]a, b]$ if and only if g is integrable on $[a', b'[$. In this way we are reconducted to the previous theory.

We will also define the integral of a function $f :]a, b[\to \mathbb{R}$, with $-\infty \le a < b \le +\infty$, in this way:

Definition 1.45

We say that $f :]a, b[\to \mathbb{R}$ is **integrable** if, once we fix a point $p \in]a, b[$, the function f is integrable on $[p, b[$ and on $]a, p]$. In that case, the **integral** of f on $]a, b[$ is defined by

$$\int_a^b f = \int_a^p f + \int_p^b f .$$

It is easy to verify that the given definition does not depend on the choice of $p \in]a, b[$.

Examples If $a, b \in \mathbb{R}$, one can verify that the function

$$f(x) = ((x - a)(b - x))^{-\beta}$$

is integrable on $]a, b[$ if and only if $\beta < 1$. In this case, it is possible to choose, for instance, $p = (a + b)/2$. Another case arises when $a = -\infty$ and $b = +\infty$. For example, one easily verifies that the function $f(x) = (x^2+1)^{-1}$ is integrable on $]-\infty, +\infty[$. Taking for instance $p = 0$, we have:

$$\int_{-\infty}^{+\infty} \frac{1}{x^2 + 1}\,dx = \int_{-\infty}^{0} \frac{1}{x^2 + 1}\,dx + \int_{0}^{+\infty} \frac{1}{x^2 + 1}\,dx = \pi .$$

A further case one could face in the applications is when a function happens not to be defined in an interior point of an interval.

Definition 1.46
Given $a < q < b$, we say that a function $f : [a, b] \setminus \{q\} \to \mathbb{R}$ is integrable if f is both integrable on $[a, q[$ and on $]q, b]$. In that case, we set

$$\int_a^b f = \int_a^q f + \int_q^b f .$$

For example, if $a < 0 < b$, the function $f(x) = \sqrt{|x|}/x$ is integrable on $[a, b] \setminus \{0\}$, and

$$\int_a^b \frac{\sqrt{|x|}}{x}\,dx = \int_a^0 \frac{-1}{\sqrt{-x}}\,dx + \int_0^b \frac{1}{\sqrt{x}}\,dx = 2\sqrt{b} - 2\sqrt{-a} .$$

On the other hand, the function $f(x) = 1/x$ is not integrable on $[a, b] \setminus \{0\}$, even if the fact that f is odd could lead someone to define the integral on symmetric intervals with respect to the origin as being equal to zero. However, by doing so, some important properties of the integral would be lost, as for example the additivity on sub-intervals.

Different situations can be faced combining together those treated above. I prefer not to go deeper into these details; in each single case, the choice of the appropriate method will be made by the right guess.

1.16 The Hake Theorem

Recall that a function $f : [a, b[\to \mathbb{R}$ is said to be integrable if it is integrable on $[a, c]$, for every $c \in]a, b[$, and the limit

$$\lim_{c \to b^-} \int_a^c f$$

exists and is finite. We want to prove the following result due to H. Hake.

Theorem 1.47
Let $b < +\infty$, and assume $f : [a, b[\to \mathbb{R}$ to be a function which is integrable on $[a, c]$, for every $c \in]a, b[$. Then, the function f is integrable on $[a, b[$ if and only if it is the restriction of an integrable function $\bar{f} : [a, b] \to \mathbb{R}$. In that case,

$$\int_a^b \bar{f} = \int_a^b f .$$

Proof

Assume first that f be the restriction of an integrable function $\bar{f} : [a,b] \to \mathbb{R}$. Fix $\varepsilon > 0$; we want to find a $\gamma > 0$ such that, if $c \in]a,b[$ and $b - c \le \gamma$, then

$$\left| \int_a^c f - \int_a^b \bar{f} \right| \le \varepsilon .$$

Let δ be a gauge such that, for every δ-fine P-partition of $[a,b]$, it is $|S(I, \bar{f}, \Pi) - \int_a^b \bar{f}| \le \frac{\varepsilon}{8}$. We choose a positive constant $\gamma \le \delta(b)$ such that $\gamma|\bar{f}(b)| \le \frac{\varepsilon}{2}$. If $c \in]a,b[$ and $b - c \le \gamma$, by the Saks–Henstock theorem, taking the δ-fine sub-P-partition $\Pi = \{(b, [c,b])\}$, we have

$$\left| \bar{f}(b)(b-c) - \int_c^b \bar{f} \right| \le 4 \frac{\varepsilon}{8} = \frac{\varepsilon}{2} ,$$

and hence

$$\left| \int_a^c f - \int_a^b \bar{f} \right| = \left| \int_c^b \bar{f} \right| \le \frac{\varepsilon}{2} + |\bar{f}(b)(b-c)| \le \frac{\varepsilon}{2} + |\bar{f}(b)|\gamma \le \frac{\varepsilon}{2} + \frac{\varepsilon}{2} = \varepsilon .$$

Let us prove now the other implication. Assume f to be integrable on $[a,b[$, and let A be its integral. We extend f to a function $\bar{f}$ defined on the whole interval $[a,b]$, by setting, for instance, $\bar{f}(b) = 0$. In order to prove that $\bar{f}$ is integrable on $[a,b]$ with integral A, fix $\varepsilon > 0$. There is a $\gamma > 0$ such that, if $c \in]a,b[$ and $b - c \le \gamma$, then

$$\left| \int_a^c f - A \right| \le \frac{\varepsilon}{2} .$$

Consider the sequence $(c_i)_i$ of points in $[a,b[$, given by

$$c_i = b - \frac{b-a}{i+1} .$$

Notice that it is strictly increasing, it converges to b, and it is $c_0 = a$. Since f is integrable on each interval $[c_{i-1}, c_i]$, we can consider, for each $i \ge 1$, a gauge δ_i on $[c_{i-1}, c_i]$ such that, for every δ_i-fine P-partition Π_i of $[c_{i-1}, c_i]$, one has

$$\left| S([c_{i-1}, c_i], f, \Pi_i) - \int_{c_{i-1}}^{c_i} f \right| \le \frac{\varepsilon}{2^{i+4}} .$$

We define a gauge δ on $[a,b]$ by setting

$$\delta(x) = \begin{cases} \min\left\{\delta_i(x), \frac{x-c_{i-1}}{2}, \frac{c_i - x}{2}\right\} & \text{if } x \in]c_{i-1}, c_i[, \\ \min\left\{\delta_1(a), \frac{c_1 - a}{2}\right\} & \text{if } x = a , \\ \min\left\{\delta_i(c_i), \delta_{i+1}(c_i), \frac{c_i - c_{i-1}}{2}, \frac{c_{i+1} - c_i}{2}\right\} & \text{if } x = c_i \text{ and } i \ge 1 , \\ \gamma & \text{if } x = b . \end{cases}$$

Let $\Pi = \{(x_j, [a_{j-1}, a_j]) : j = 1, \ldots, m\}$, be a δ-fine P-partition of $[a, b]$. Denote by q be the smallest integer for which $c_{q+1} \geq a_{m-1}$. The choice of the gauge permits to split the Riemann sum, similarly as in the proof of the theorem on the additivity of the integral on subintervals. The sum $S([a, b], \bar{f}, \Pi)$ will thus contain

- q Riemann sums on $[c_{i-1}, c_i]$, with $i = 1, \ldots, q$;
- a Riemann on $[c_q, a_{m-1}]$;
- a last term $\bar{f}(x_m)(b - a_{m-1})$.

To better clarify what we just said, assume for example that $q = 2$; then there must be a $\bar{j}_1$ for which $x_{\bar{j}_1} = c_1$, and a $\bar{j}_2$ for which $x_{\bar{j}_2} = c_2$. Then,

$$
\begin{aligned}
S([a, b], \bar{f}, \Pi) = {} & [f(x_1)(a_1 - a) + \ldots + f(x_{\bar{j}_1 - 1})(a_{\bar{j}_1 - 1} - a_{\bar{j}_1 - 2}) + f(c_1)(c_1 - a_{\bar{j}_1 - 1})] \\
& + [f(c_1)(a_{\bar{j}_1} - c_1) + \ldots + f(x_{\bar{j}_2 - 1})(a_{\bar{j}_2 - 1} - a_{\bar{j}_2 - 2}) + f(c_2)(c_2 - a_{\bar{j}_2 - 1})] \\
& + [f(c_2)(a_{\bar{j}_2} - c_2) + \cdots + f(x_{m-1})(a_{m-1} - a_{m-2})] \\
& + \bar{f}(x_m)(b - a_{m-1}) .
\end{aligned}
$$

In general, we will have

$$
\begin{aligned}
S([a, b], \bar{f}, \Pi) = \sum_{i=1}^{q} S([c_{i-1}, c_i], f, \Pi_i) + S([c_q, a_{m-1}], f, \Pi_{q+1}) + \\
+ \bar{f}(x_m)(b - a_{m-1}) ,
\end{aligned}
$$

where, for $i = 1, \ldots, q$, Π_i is a δ_i-fine P-partition of $[c_{i-1}, c_i]$, while Π_{q+1} is a δ_{q+1}-fine P-partition of $[c_q, a_{m-1}]$, and hence a δ_{q+1}-fine sub-P-partition of $[c_q, c_{q+1}]$. By the choice of the gauge δ, it has to be $x_m = b$ and hence $\bar{f}(x_m) = 0$. Moreover, since $\delta(x_m) = \gamma$, it is $b - a_{m-1} \leq \gamma$. Using the fact that

$$
\int_a^{a_{m-1}} f = \sum_{i=1}^{q} \int_{c_{i-1}}^{c_i} f + \int_{c_q}^{a_{m-1}} f ,
$$

by the Saks–Henstock theorem we have

$$
\begin{aligned}
|S([a, b], \bar{f}, \Pi) - A| & \leq \left| S([a, b], \bar{f}, \Pi) - \int_a^{a_{m-1}} f \right| + \left| \int_a^{a_{m-1}} f - A \right| \\
\leq \sum_{i=1}^{q} & \left| S([c_{i-1}, c_i], f, \Pi_i) - \int_{c_{i-1}}^{c_i} f \right| + \\
& + \left| S([c_q, a_{m-1}], f, \Pi_{q+1}) - \int_{c_q}^{a_{m-1}} f \right| + \left| \int_a^{a_{m-1}} f - A \right|
\end{aligned}
$$

$$\leq \sum_{i=1}^{q} \frac{\varepsilon}{2^{i+4}} + 4\,\frac{\varepsilon}{2^{q+4}} + \frac{\varepsilon}{2}$$

$$\leq \frac{\varepsilon}{4} + \frac{\varepsilon}{4} + \frac{\varepsilon}{2} = \varepsilon\,,$$

and the proof is thus completed. □

The above theorem suggests that even for a function of the type $f : [a, +\infty[\to \mathbb{R}$ the definition of the improper integral could be reduced to that of a usual integral. Indeed, fixing arbitrarily $b > a$, we could define a continuously differentiable strictly increasing function $\varphi : [a, b[\to \mathbb{R}$ such that $\varphi(a) = a$ and $\lim_{u \to b^-} \varphi(u) = +\infty$; for example, take $\varphi(u) = a + \ln \frac{b-a}{b-u}$. A formal change of variables then gives

$$\int_a^{+\infty} f(x)\,dx = \int_a^b f(\varphi(u))\varphi'(u)\,du\,,$$

and to this last integral Hake's theorem applies.

With this idea in mind, it is possible to prove that $f : [a, +\infty[\to \mathbb{R}$ is integrable and its integral is a real number A if and only if for every $\varepsilon > 0$ there is a gauge δ, defined on $[a, +\infty[$, and a positive constant α such that, if

$$a = a_0 < a_1 < \cdots < a_{m-1}\,, \quad \text{with} \quad a_{m-1} \geq \alpha\,,$$

and, for every $j = 1, \ldots, m-1$,

$$x_j - \delta(x_j) \leq a_{j-1} \leq x_j \leq a_j \leq x_j + \delta(x_j)\,,$$

then

$$\left| \sum_{j=1}^{m-1} f(x_j)(a_j - a_{j-1}) - A \right| \leq \varepsilon\,.$$

We refer to the book of Bartle [1] for a complete treatment of this case.

Needless to say, similar considerations can be made in the case when the function f is defined on an interval of the type $]a, b]$, with $-\infty \leq a$.

1.17 Integrals and Series

We now prove a theorem which shows the close connection between the theory of numerical series and that of the improper integral.

Theorem 1.48
Let $f : [1, +\infty[\to \mathbb{R}$ *be a function which is positive, decreasing and integrable on* $[1, c]$, *for every* $c > 1$. *Then* f *is integrable on* $[1, +\infty[$ *if and only if the series* $\sum_{k=1}^{\infty} f(k)$ *converges. Moreover, we have*

$$\sum_{k=2}^{\infty} f(k) \leq \int_1^{+\infty} f \leq \sum_{k=1}^{\infty} f(k)\,.$$

Proof
For $x \in [k, k+1]$, it has to be $f(k+1) \leq f(x) \leq f(k)$. Hence,

$$f(k+1) \leq \int_k^{k+1} f \leq f(k)\,.$$

Summing up, we obtain

$$\sum_{k=1}^{n} f(k+1) \leq \int_1^{n+1} f \leq \sum_{k=1}^{n} f(k)\,.$$

Being f positive, the sequence $(\sum_{k=1}^{n} f(k))_n$ and the function $c \mapsto \int_1^c f$ are both increasing and therefore have a limit. The conclusion now follows from the comparison theorem for limits. □

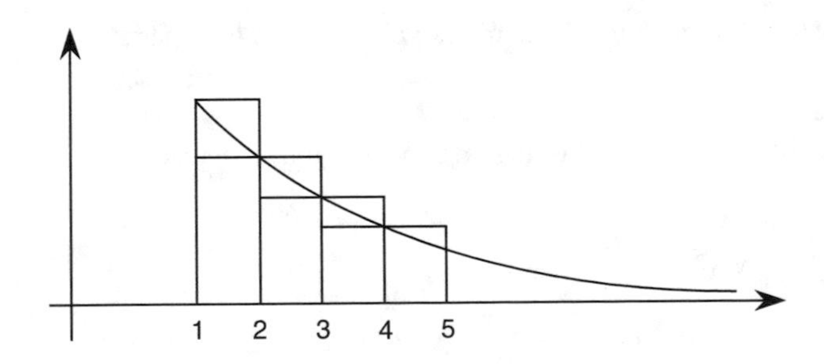

Observations It is clear that the choice $a = 1$ in the theorem just proved is by no way necessary. Notice moreover that this theorem is often used to determine the convergence of a series, giving the estimate

$$\int_1^{+\infty} f \leq \sum_{k=1}^{\infty} f(k) \leq f(1) + \int_1^{+\infty} f\,.$$

Example Consider the series $\sum_{k=1}^{\infty} k^{-3}$; in this case:

$$\int_1^{+\infty} \frac{1}{x^3}\, dx \le \sum_{k=1}^{\infty} \frac{1}{k^3} \le 1 + \int_1^{+\infty} \frac{1}{x^3}\, dx\,,$$

and then

$$\frac{1}{2} \le \sum_{k=1}^{\infty} \frac{1}{k^3} \le \frac{3}{2}\,.$$

A greater accuracy is easily attained by computing the sum of a few first terms and then using the estimate given by the integral. For example, separating the first two terms, we have:

$$\sum_{k=1}^{\infty} \frac{1}{k^3} = 1 + \frac{1}{8} + \sum_{k=3}^{\infty} \frac{1}{k^3}\,,$$

with

$$\int_3^{+\infty} \frac{1}{x^3}\, dx \le \sum_{k=3}^{\infty} \frac{1}{k^3} \le \frac{1}{27} + \int_3^{+\infty} \frac{1}{x^3}\, dx\,.$$

We thus have the following estimate:

$$\frac{255}{216} \le \sum_{k=1}^{\infty} \frac{1}{k^3} \le \frac{263}{216}\,.$$

We conclude this chapter with some exercises on this final part.

Exercises

1. Establish whether the following improper integrals converge:

$$\int_0^{+\infty} \sqrt{\frac{x}{x^3+1}}\, dx$$

$$\int_0^{+\infty} \frac{1}{\sqrt{x}\,(1+\sqrt{x}+x)}\, dx$$

2. For which $\alpha \in \mathbb{R}$ the integrals

$$\int_\pi^{+\infty} \frac{1}{x(\ln x)^\alpha}\, dx\,, \qquad \int_\pi^{+\infty} \frac{1}{x(\ln x)(\ln(\ln x))^\alpha}\, dx$$

converge?

3. The following improper integrals converge?

$$\int_0^2 \frac{\ln x}{\sqrt{x}\cos x}\,dx\,, \qquad \int_0^1 \frac{1}{x\ln x}\,dx$$

4. The following series converge?

$$\sum_{k=1}^{\infty} \frac{1}{k\,\ln k}\,, \qquad \sum_{k=1}^{\infty} \frac{1}{k\,\ln k\,\ln(\ln k)}$$

Functions of Several Real Variables

Alessandro Fonda

A. Fonda, *The Kurzweil-Henstock Integral for Undergraduates*,
Compact Textbooks in Mathematics,
https://doi.org/10.1007/978-3-319-95321-2_2

In this chapter we extend the theory developed in the previous one to functions of several variables defined on subsets of $\mathbb{R}^N$, with values in $\mathbb{R}$. In order to simplify the exposition, we will often concentrate on the case $N = 2$. It will not be so difficult for the reader to extend the various results to the case of a generic dimension N.

2.1 Integrability on Rectangles

We begin by considering the case of functions defined on rectangles. A **rectangle** of $\mathbb{R}^N$ is a set of the type $[a_1, b_1] \times \cdots \times [a_N, b_N]$. This word is surely familiar in the case $N = 2$. If $N = 1$, a rectangle happens to be a compact interval while, if $N = 3$, usually one prefers to call it "rectangle parallelepiped". In the following exposition, we concentrate for simplicity on the two-dimensional case. The general case is perfectly similar and does not involve greater difficulties, except for the notations.

We consider the rectangle $I = [a_1, b_1] \times [a_2, b_2] \subseteq \mathbb{R}^2$. Let us define the measure of I :

$$\mu(I) = (b_1 - a_1)(b_2 - a_2) .$$

We say that two rectangles are non-overlapping if their interiors are disjoint.

A **P-partition** of the rectangle I is a set

$$\Pi = \{(\boldsymbol{x}_1, I_1), (\boldsymbol{x}_2, I_2), \dots, (\boldsymbol{x}_m, I_m)\} ,$$

where the I_j are non-overlapping rectangles whose union is I and, for every $j = 1, \dots, m$, the point $\boldsymbol{x}_j = (x_j, y_j)$ belongs to I_j.

Example If $I = [0, 10] \times [0, 6]$, a possible P-partition is the following:

$$\Pi = \{((1,1), [0,7] \times [0,2]), ((0,5), [0,3] \times [2,6]),$$
$$((5,4), [3,10] \times [4,6]), ((10,0), [7,10] \times [0,4]), ((5,3), [3,7] \times [2,4])\}.$$

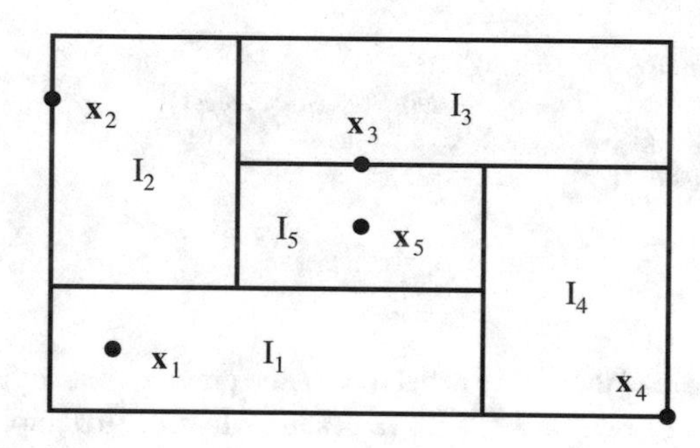

Let us now consider a function f defined on the rectangle I, with values in $\mathbb{R}$, and let $\Pi = \{(\boldsymbol{x}_j, I_j) : j = 1, \dots, m\}$ be a P-partition of I. We call **Riemann sum** associated to I, f and Π the real number $S(I, f, \Pi)$ defined by

$$S(I, f, \Pi) = \sum_{j=1}^{m} f(\boldsymbol{x}_j)\mu(I_j).$$

Whenever f happens to be positive, this number is the sum of the volumes of the parallelepipeds having as base I_j and height $[0, f(\boldsymbol{x}_j)]$.

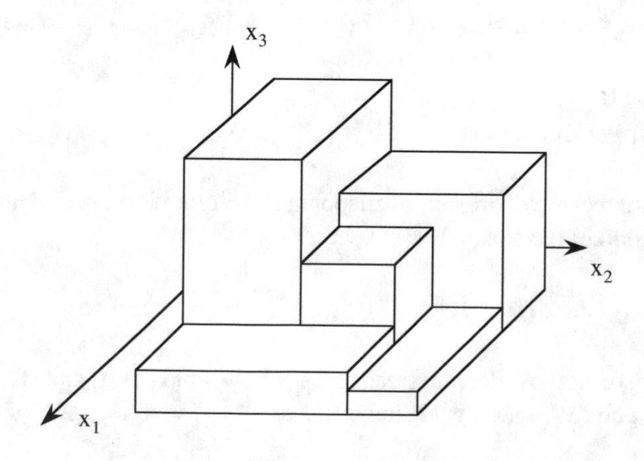

We now introduce the notion of fineness for the P-partition Π defined above. We call **gauge** on I every function $\delta : I \to \mathbb{R}$ such that $\delta(\boldsymbol{x}) > 0$ for every $\boldsymbol{x} \in I$. Given a gauge δ on I, we say that the P-partition Π introduced above is **δ-fine** if, for every $j = 1, \dots, m$,

$$I_j \subseteq [x_j - \delta(x_j, y_j), x_j + \delta(x_j, y_j)] \times [y_j - \delta(x_j, y_j), y_j + \delta(x_j, y_j)] .$$

In the following, given $\boldsymbol{x} = (x, y) \in I$ and $r > 0$, in order to shorten the notations we will write

$$B[\boldsymbol{x}, r] = [x - r, x + r] \times [y - r, y + r] ;$$

the P-partition Π will then be δ-fine if, for every $j = 1, \dots, m$,

$$I_j \subseteq B[\boldsymbol{x}_j, \delta(\boldsymbol{x}_j)] .$$

Example Let $I = [0, 1] \times [0, 1]$ and δ be the gauge defined as follows:

$$\delta(x, y) = \begin{cases} \dfrac{x+y}{3} & \text{if } (x, y) \neq (0, 0) , \\ \dfrac{1}{2} & \text{if } (x, y) = (0, 0) . \end{cases}$$

We want to find a δ-fine P-partition of I. Similarly as was seen in the case $N = 1$ we have in this case that one of the points $\boldsymbol{x}_j$ necessarily has to coincide with $(0, 0)$. We can then choose, for example,

$$\Pi = \left\{ \left((0,0), \left[0, \frac{1}{2}\right] \times \left[0, \frac{1}{2}\right] \right), \left(\left(\frac{1}{2}, 1\right), \left[0, \frac{1}{2}\right] \times \left[\frac{1}{2}, 1\right] \right), \right.$$
$$\left. \left(\left(1, \frac{1}{2}\right), \left[\frac{1}{2}, 1\right] \times \left[0, \frac{1}{2}\right] \right), \left((1,1), \left[\frac{1}{2}, 1\right] \times \left[\frac{1}{2}, 1\right] \right) \right\} .$$

It is interesting to observe that it is not always possible to construct δ-fine P-partitions by only joining points on the edges of I. The reader can be convinced by trying to do this with the following gauge:

$$\delta(x, y) = \begin{cases} \dfrac{x+y}{16} & \text{if } (x, y) \neq (0, 0) , \\ \dfrac{1}{2} & \text{if } (x, y) = (0, 0) . \end{cases}$$

As in the one-dimensional case, one can prove that for every gauge δ on I there exists a δ-fine P-partition of I (Cousin's Theorem). The following definition is identical to the one seen in Chapter 1.

Definition 2.1

A function $f : I \to \mathbb{R}$ is said to be **integrable** (on the rectangle I) if there is a real number A with the following property: given $\varepsilon > 0$, it is possible to find a gauge δ on I such that, for every δ-fine P-partition Π of I, it is

$$|S(I, f, \Pi) - A| \leq \varepsilon .$$

We briefly overview all the properties which can be obtained from the given definition in the same way as was done in the case of a function of a single variable.

There is at most one $A \in \mathbb{R}$ which verifies the conditions of the definition. Such a number is called the **integral** of f on I and is denoted by one of the following symbols:

$$\int_I f\,, \qquad \int_I f(\boldsymbol{x})\,d\boldsymbol{x}\,, \qquad \int_I f(x, y)\,dx\,dy\,.$$

The set of integrable functions is a vector space, and the integral is a linear function on it:

$$\int_I (f + g) = \int_I f + \int_I g\,, \qquad \int_I (\alpha f) = \alpha \int_I f$$

(with $\alpha \in \mathbb{R}$); it preserves the order:

$$f \leq g \quad \Rightarrow \quad \int_I f \leq \int_I g\,.$$

The **Cauchy criterion of integrability** holds. Moreover, we have the following version of the theorem on **additivity on subrectangles**.

Theorem 2.2

Let $f : I \to \mathbb{R}$ be a function and $K_1, K_2, \dots, K_l$ be non-overlapping sub-rectangles of I whose union is I. Then, f is integrable on I if and only if it is integrable on each of the K_i. In that case, we have

$$\int_I f = \sum_{i=1}^{l} \int_{K_i} f.$$

In particular, if a function is integrable on a rectangle, it still is on every subrectangle. The proof of the theorem is similar to that given in the one-dimensional case, and is based on the possibility of constructing a gauge which permits to split the Riemann sums on I in the sum of Riemann sums on the single subrectangles.

We say that an integrable function on I is **R-integrable** there (or integrable according to Riemann) if, among all possible gauges $\delta : I \to \mathbb{R}$ which verify the definition of integrability, it is always possible to choose one which is constant on I. The set of R-integrable functions is a vector subspace of the space of integrable functions and contains the subspace of continuous functions.

We say that an integrable function $f : I \to \mathbb{R}$ is **L-integrable** (or integrable according to Lebesgue) if $|f|$ is integrable on I, as well. The L-integrable functions make up a vector subspace of the space of integrable functions. If f and g are two L-integrable functions on I, then the functions $\min\{f, g\}$ and $\max\{f, g\}$ are L-integrable on I, too. A function f is L-integrable on I if and only if such are its positive part $f^+ = \max\{f, 0\}$ and its negative part $f^- = \max\{-f, 0\}$.

The **Saks–Henstock's theorem**, the **Monotone Convergence Theorem** of B. Levi and the **Dominated Convergence Theorem** of H. Lebesgue extend, with statements and proofs perfectly analogous to those given in the first chapter, to the integrable functions on a rectangle.

Exercises

1. Let $f : [a, b] \times [c, d] \to \mathbb{R}$ be defined as $f(x, y) = x$. Prove that f is integrable and compute

$$\int_{[a,b]\times[c,d]} f(x, y)\, dx\, dy\,.$$

2. Let $f : [0, 1] \times [0, 1] \to \mathbb{R}$ be the function defined as

$$f(x, y) = \begin{cases} 3 & \text{if } x \in \left[0, \frac{1}{2}\right], \\ 5 & \text{if } x \in \left[\frac{1}{2}, 1\right]. \end{cases}$$

Prove that f is integrable with

$$\int_{[0,1]\times[0,1]} f(x, y)\, dx\, dy = 4\,.$$

3. Prove that, if $f : I \to \mathbb{R}$ is equal to 0 for every $(x, y) \notin \mathbb{Q} \times \mathbb{Q}$, then $\int_I f = 0$.
4. By the use of the dDominated Convergence Theorem, prove that

$$\lim_{k\to\infty} \int_{[0,1]\times[0,1]} \frac{\cos(e^{k(x^2+y^2)})}{\sqrt{k}}\, dx\, dy = 0\,,$$

$$\lim_{k\to\infty} \int_{[-1,1]\times[-1,1]} \sqrt{k}\, \sinh\left(\frac{x+y}{k}\right) dx\, dy = 0\,.$$

2.2 Integrability on a Bounded Set

Given a bounded set E and a function f whose domain contains E, we define the function f_E as follows:

$$f_E(\boldsymbol{x}) = \begin{cases} f(\boldsymbol{x}) & \text{if } \boldsymbol{x} \in E\,, \\ 0 & \text{if } \boldsymbol{x} \notin E\,. \end{cases}$$

We can prove the following

Proposition 2.3
Let I_1 and I_2 be two rectangles containing the set E. Then, f_E is integrable on I_1 if and only if it is integrable on I_2. In that case, we have $\int_{I_1} f_E = \int_{I_2} f_E$.

Proof
We consider for simplicity the case $N = 2$. Assume that f_E be integrable on I_1. Let K be a rectangle containing both I_1 and I_2. We can construct some non-overlapping rectangles $K_1, \ldots, K_r$, also non-overlapping with I_1, such that $I_1 \cup K_1 \cup \cdots \cup K_r = K$. We now prove that f_E is integrable on each of the subrectangles $K_1, \ldots, K_r$, and that the integrals $\int_{K_1} f_E, \ldots, \int_{K_r} f_E$ are all equal to zero. Notice that f_E restricted to each of these subrectangles is zero everywhere except perhaps on one of their edges. We are thus led to prove the following lemma, which will permit us to conclude the proof.

Lemma 2.4
Let $\mathcal{K}$ be a rectangle and $g : \mathcal{K} \to \mathbb{R}$ be a function which is zero everywhere except perhaps on one edge of $\mathcal{K}$. Then g is integrable on $\mathcal{K}$ and $\int_{\mathcal{K}} g = 0$.

Proof
We first assume that the function g be bounded on $\mathcal{K}$, i.e., that there is a constant $C > 0$ for which

$$|g(x, y)| \leq C\,,$$

for every $(x, y) \in \mathcal{K}$. Fix $\varepsilon > 0$. Let L be the edge of the rectangle $\mathcal{K}$ on which g can be nonzero, and denote by ℓ its length. Define the constant gauge $\delta = \frac{\varepsilon}{C\ell}$. Then, for every δ-fine

P-partition $\Pi = \{(\boldsymbol{x}_1, I_1), (\boldsymbol{x}_2, I_2), \dots, (\boldsymbol{x}_m, I_m)\}$ of $\mathcal{K}$, we have:

$$\begin{aligned}|S(\mathcal{K}, g, \Pi)| &\le \sum_{j=1}^{m} |g(\boldsymbol{x}_j)|\mu(I_j)\\ &= \sum_{\{j\,:\,\boldsymbol{x}_j \in L\}} |g(\boldsymbol{x}_j)|\mu(I_j)\\ &\le C \sum_{\{j\,:\,\boldsymbol{x}_j \in L\}} \mu(I_j)\\ &\le C\delta\ell = \varepsilon\,.\end{aligned}$$

This proves that g is integrable on $\mathcal{K}$ and $\int_{\mathcal{K}} g = 0$ in the case when g is bounded on $\mathcal{K}$. If g is not such, assume that it has non-negative values. Define the following sequence $(g_k)_k$ of functions:

$$g_k(\boldsymbol{x}) = \min\{g(\boldsymbol{x}), k\}\,.$$

Being the functions g_k bounded, for what have been seen above we have $\int_{\mathcal{K}} g_k = 0$, for every k. It is easily seen that the sequence thus defined satisfies the conditions of the Monotone Convergence Theorem and converges pointwise to g. It then follows that g is integrable on $\mathcal{K}$ and

$$\int_{\mathcal{K}} g = \lim_k \int_{\mathcal{K}} g_k = 0\,.$$

If g does not have only non-negative values, it is always possible to consider g^+ and g^-. From the above, $\int_{\mathcal{K}} g^+ = \int_{\mathcal{K}} g^- = 0$, and then $\int_{\mathcal{K}} g = 0$, which is what was to be proved. □

End of the Proof

Having proved that f_E is integrable on each of the $K_1, \dots, K_r$ and that the integrals $\int_{K_1} f_E, \dots, \int_{K_r} f_E$ are equal to zero, by the theorem of additivity on subrectangles we have that, being f_E integrable on I_1, it is such on K, and

$$\int_K f_E = \int_{I_1} f_E + \int_{K_1} f_E + \dots + \int_{K_r} f_E = \int_{I_1} f_E\,.$$

But then f_E is integrable on every sub-interval of K, and in particular on I_2. We can now construct, analogously to what has just been done for I_1, some non-overlapping rectangles $J_1, \dots, J_s$, also non-overlapping with I_2, such that $I_2 \cup J_1 \cup \dots \cup J_s = K$. Similarly, we will have

$$\int_K f_E = \int_{I_2} f_E + \int_{J_1} f_E + \dots + \int_{J_s} f_E = \int_{I_2} f_E\,,$$

which proves that $\int_{I_1} f_E = \int_{I_2} f_E$. To see that the condition is necessary and sufficient, just exchange the roles of I_1 and I_2 in the above proof. □

We are thus led to the following.

Definition 2.5

Given a bounded set E, we say that the function $f : E \to \mathbb{R}$ is **integrable** (on the set E) if there is a rectangle I containing the set E on which f_E is integrable. In that case, we set

$$\int_E f = \int_I f_E \,.$$

When f is integrable on E according to the previous definition, one has that f_E is integrable on any rectangle containing the set E, and the integral of f_E remains the same on each such rectangle.

With the given definition, **all the properties of the integral seen before easily extend**. *There is an exception concerning the additivity*, since it is not true in general that a function which is integrable on a bounded set remains integrable on any of its subsets. Indeed, take a function $f : E \to \mathbb{R}$ which is integrable but not L-integrable. We consider the subset

$$E' = \{x \in E : f(x) \geq 0\} \,,$$

and we show that f cannot be integrable on E'. If it was, then f^+ would be integrable on E. But then also $f^- = f^+ - f$ would be integrable on E, and therefore f should be L-integrable on E, in contradiction with the assumption.

We will see that, with respect to additivity, the L-integrable functions have a somewhat better behavior.

2.3 The Measure of a Bounded Set

Definition 2.6

A bounded set E is said to be **measurable** if the constant function 1 is integrable on E. In that case, the number $\int_E 1$ is said to be the **measure** of E and is denoted by $\mu(E)$.

The measure of a measurable set is thus a non-negative number. The empty set is assumed to be measurable, and its measure is equal to 0. In the case of a subset of $\mathbb{R}^2$, its measure is also called the **area** of the set. If $E = [a_1, b_1] \times [a_2, b_2]$ is a rectangle, it is easily seen that

$$\mu(E) = \int_E 1 = (b_1 - a_1)(b_2 - a_2) \,,$$

so that the notation is in accordance with the one already introduced for rectangles. For a subset of $\mathbb{R}^3$, the measure is also called the **volume** of the set.

Not every set is measurable. It is shown in Appendix C that, when dealing with non-measurable sets, some paradoxical situations can arise. In the following, we will be careful to always consider measurable sets.

Let us analyze some properties of the measure. It is useful to introduce the characteristic function of a set E, defined by

$$\chi_E(\boldsymbol{x}) = \begin{cases} 1 & \text{if } \boldsymbol{x} \in E\,, \\ 0 & \text{if } \boldsymbol{x} \notin E\,. \end{cases}$$

If I is a rectangle containing the set E, we thus have

$$\mu(E) = \int_I \chi_E\,.$$

Proposition 2.7
Let A and B be two bounded and measurable sets. The following properties hold:
(*a*) *if $A \subseteq B$, then $B \setminus A$ is measurable, and*

$$\mu(B \setminus A) = \mu(B) - \mu(A)\,;$$

in particular, $\mu(A) \leq \mu(B)$.
(*b*) *$A \cup B$ and $A \cap B$ are measurable, and*

$$\mu(A \cup B) + \mu(A \cap B) = \mu(A) + \mu(B)\,;$$

in particular, if A and B are disjoint, then $\mu(A \cup B) = \mu(A) + \mu(B)$.

Proof
Let I be a rectangle containing $A \cup B$. If $A \subseteq B$, then $\chi_{B\setminus A} = \chi_B - \chi_A$, and property (*a*) follows by integrating on I.

Being $\chi_{A\cup B} = \max\{\chi_A, \chi_B\}$ and $\chi_{A\cap B} = \min\{\chi_A, \chi_B\}$, we have that $\chi_{A\cup B}$ and $\chi_{A\cap B}$ are integrable on I. Moreover,

$$\chi_{A\cup B} + \chi_{A\cap B} = \chi_A + \chi_B\,,$$

and integrating on I we have (*b*). □

The following proposition states the property of **complete additivity** of the measure.

Proposition 2.8
If $(A_k)_{k\geq 1}$ is a sequence of bounded and measurable sets, whose union $A = \bigcup_{k\geq 1} A_k$ is bounded, then A is measurable, and

$$\mu(A) \leq \sum_{k=1}^{\infty} \mu(A_k)\,.$$

If the sets A_k are pairwise disjoint, then equality holds.

Proof
Assume first that the sets A_k be pairwise disjoint. Let I be a rectangle containing their union A. Then, for every $\boldsymbol{x} \in I$,

$$\chi_A(\boldsymbol{x}) = \sum_{k=1}^{\infty} \chi_{A_k}(\boldsymbol{x})\,.$$

Moreover, since for every positive integer q it is

$$\sum_{k=1}^{q} \mu(A_k) = \mu\left(\bigcup_{k=1}^{q} A_k\right) \leq \mu(I)\,,$$

the series $\sum_{k=1}^{\infty} \int_I \chi_{A_k} = \sum_{k=1}^{\infty} \mu(A_k)$ converges. By the corollary following the Monotone Convergence Theorem, we have that A is measurable and

$$\mu(A) = \int_I \chi_A = \int_I \sum_{k=1}^{\infty} \chi_{A_k} = \sum_{k=1}^{\infty} \int_I \chi_{A_k} = \sum_{k=1}^{\infty} \mu(A_k)\,.$$

When the sets A_k are not pairwise disjoint, consider the sets $B_1 = A_1$, $B_2 = A_2 \setminus A_1$ and, in general, $B_k = A_k \setminus (A_1 \cup \cdots \cup A_{k-1})$. The sets B_k are measurable, pairwise disjoint, and $\bigcup_{k\geq 1} B_k = \bigcup_{k\geq 1} A_k$. The conclusion then follows from what has been proved above. □

We have a similar proposition concerning the intersection of a countable family of sets.

Proposition 2.9
If $(A_k)_{k\geq 1}$ is a sequence of bounded and measurable sets, their intersection $A = \bigcap_{k\geq 1} A_k$ is a measurable set.

Proof
Let I be a rectangle containing the set A. Then,

$$\bigcap_{k\geq 1} A_k = I \setminus \left(\bigcup_{k\geq 1} (I \setminus (A_k \cap I)) \right),$$

and the conclusion follows from the two previous propositions. □

The following two propositions will provide us with a large class of measurable sets.

Proposition 2.10
Every open and bounded set is measurable.

Proof
Consider for simplicity the case $N = 2$. Let A be an open set contained in a rectangle I. We divide the rectangle I in four rectangles of equal areas using the axes of its edges. Then we proceed analogously with each of these four rectangles, thus obtaining sixteen smaller rectangles, and so on. Being A open, for every $\boldsymbol{x} \in A$ there is a small rectangle among those just constructed which contains $\boldsymbol{x}$ and is contained in A. In this way, it is seen that the set A is covered by a countable family of rectangles; being the union of a countable family of measurable sets, it is therefore measurable. □

Proposition 2.11
Every compact set is measurable.

Proof
Let B be a compact set, and let I be a rectangle whose interior $\mathring{I}$ contains B. Being $\mathring{I}$ and $\mathring{I} \setminus B$ open and hence measurable, we have that $B = \mathring{I} \setminus (\mathring{I} \setminus B)$ is measurable. □

Example The set

$$E = \{(x, y) \in \mathbb{R}^2 : 1 < x^2 + y^2 \leq 4\}$$

is measurable, being the difference of the closed disks with radius 2 and 1 centered at the origin:

$$E = \{(x, y) \in \mathbb{R}^2 : x^2 + y^2 \leq 4\} \setminus \{(x, y) \in \mathbb{R}^2 : x^2 + y^2 \leq 1\}.$$

2

2.4 The Chebyshev Inequality

Theorem 2.12

Let E be a bounded set, and $f : E \to \mathbb{R}$ be an integrable function, with non-negative values. Then, for every $r > 0$, the set

$$E_r = \{\boldsymbol{x} \in E : f(\boldsymbol{x}) > r\}$$

is measurable, and

$$\mu(E_r) \leq \frac{1}{r} \int_E f .$$

Proof

Let I be a rectangle containing E. Once fixed $r > 0$, we define the following functions on I :

$$f_k(\boldsymbol{x}) = \min\{1, k \max\{f_E(\boldsymbol{x}) - r, 0\}\} .$$

They make up an increasing sequence of L-integrable functions which pointwise converges to χ_{E_r}. Clearly,

$$0 \leq f_k(\boldsymbol{x}) \leq 1 ,$$

for every k and every $\boldsymbol{x} \in I$. The Monotone (or Dominated, as well) Convergence Theorem guarantees that χ_{E_r} is integrable on I, i.e., that E_r is measurable. Since, for every $\boldsymbol{x} \in E$, it is $r\chi_{E_r}(\boldsymbol{x}) \leq f(\boldsymbol{x})$, integrating we obtain the inequality we are looking for. □

Corollary 2.13

Let E be a bounded and measurable set, and $f : E \to \mathbb{R}$ an integrable function with non-negative values. Then, taken two real numbers r, s such that $0 \leq r < s$, the set

$$E_{r,s} = \{\boldsymbol{x} \in E : r \leq f(\boldsymbol{x}) < s\}$$

is measurable.

Proof

Let $E'_r = \{x \in E : f(x) \geq r\}$. With the notations of the previous theorem, if $r > 0$, we have

$$E'_r = \bigcap_{k > \frac{1}{r}} E_{r-\frac{1}{k}} ,$$

while, if $r = 0$, we have $E'_r = E$. In any case, E'_r is measurable. Being $E_{r,s} = E'_r \setminus E'_s$, the conclusion follows. □

2.5 Negligible Sets

Definition 2.14

We say that a bounded set is **negligible** if it is measurable and its measure is equal to zero.

Every set made of a single point is negligible. Consequently, all finite or countable bounded sets are negligible. The edge of a rectangle in $\mathbb{R}^2$ is a negligible set, as shown by Lemma 2.4. Another interesting example of a negligible and not countable set is given by the *Cantor set* (see, e.g., [1, Theorem 4.16]).

By the complete additivity of the measure, the union of any sequence of negligible sets, if it is bounded, is always a negligible set.

Theorem 2.15

If E is a bounded set and $f : E \to \mathbb{R}$ is equal to zero except for a negligible set, then f is integrable on E and $\int_E f = 0$.

Proof

Let T be the negligible set on which f in non-zero. Assume first that the function f be bounded, i.e., that there is a constant $C > 0$ such that

$$|f(x)| \leq C ,$$

for every $x \in E$. We consider a rectangle I containing E and prove that $\int_I f_E = 0$. Fix $\varepsilon > 0$. Since T has zero measure, there is a gauge δ such that, for every δ-fine P-partition $\Pi = \{(x_j, I_j), j = 1, \dots, m\}$ of I,

$$S(I, \chi_T, \Pi) = \sum_{\{j : x_j \in T\}} \mu(I_j) \leq \frac{\varepsilon}{C} ,$$

so that

$$|S(I, f_E, \Pi)| \le \sum_{\{j : \boldsymbol{x}_j \in T\}} |f(\boldsymbol{x}_j)|\mu(I_j) \le C \sum_{\{j : \boldsymbol{x}_j \in T\}} \mu(I_j) \le \varepsilon .$$

Hence, in case f is bounded, it is integrable on E and $\int_E f = 0$. If f is not bounded, assume first that it has non-negative values. Define on E a sequence of functions $(f_k)_k$:

$$f_k(\boldsymbol{x}) = \min\{f(\boldsymbol{x}), k\} .$$

Being the functions f_k bounded and equal to zero except on T, for what has just been seen they are integrable on E with $\int_E f_k = 0$, for every k. It is easily seen that the defined sequence satisfies the conditions of the Monotone Convergence Theorem and converges pointwise to f. Hence, f is integrable on E, and

$$\int_E f = \lim_k \int_E f_k = 0 .$$

If f does not have non-negative values, it is sufficient to consider f^+ and f^-, and apply to them what has been said above. □

Theorem 2.16

If $f : E \to \mathbb{R}$ is an integrable function on a bounded set E, having non negative values, with $\int_E f = 0$, then f is equal to zero except on a negligible set.

Proof

Using the Chebyshev inequality, we have that, for every positive integer k,

$$\mu(E_{\frac{1}{k}}) \le k \int_E f = 0 .$$

Hence, every $E_{\frac{1}{k}}$ is negligible, and since their union is just the set where f is non-zero, we have the conclusion. □

Definition 2.17

Let E be a bounded set. We say that a proposition is true **almost everywhere** on E (or for almost every point of E) if the set of points for which it is false is negligible.

The results proved above have the following simple consequence.

Corollary 2.18
If two functions f and g, defined on the bounded set E, are equal almost everywhere on E, then f is integrable on E if and only if such is g. In that case, $\int_E f = \int_E g$.

This last corollary permits us to consider some functions which are defined almost everywhere, and to define their integral.

Definition 2.19
A function f, defined almost everywhere on E, with real values, is said to be **integrable** on E if it can be extended to an integrable function $g : E \to \mathbb{R}$. In this case, we set $\int_E f = \int_E g$.

It can be seen that **all the properties and theorems seen before remain true for such functions**. The reader is invited to verify this.

2.6 A Characterization of Bounded Measurable Sets

The following **covering lemma** will be useful in what follows.

Lemma 2.20
Let E be a set contained in a rectangle I, and let δ be a gauge on E. Then, there is a finite or countable family of non-overlapping rectangles J_k, contained in I, whose union covers the set E, with the following property: in each of the sets J_k there is a point $\boldsymbol{x}_k$ belonging to E such that $J_k \subseteq B[\boldsymbol{x}_k, \delta(\boldsymbol{x}_k)]$.

Proof
We consider for simplicity the case $N = 2$. Let us divide the rectangle I in four rectangles, having the same areas, by the axes of its edges. We proceed analogously with each of these four rectangles, obtaining sixteen smaller rectangles, and so on. We thus obtain a countable family of smaller and smaller rectangles. For every point $\boldsymbol{x}$ of E we can choose one of these rectangles which contains $\boldsymbol{x}$ and is itself contained in $B[\boldsymbol{x}, \delta(\boldsymbol{x})]$. These rectangles would satisfy the properties of the statement, if they were non-overlapping.

In order that the sets J_k be non-overlapping, it is necessary to make a careful choice of them, and here is how to do it. We first choose those from the beginning four, if there are any, which contain a point $\boldsymbol{x}_k$ belonging to E such that $J_k \subseteq B[\boldsymbol{x}_k, \delta(\boldsymbol{x}_k)]$; once this choice has been made, we eliminate all the smaller rectangles contained in them. We consider then the sixteen smaller ones and, among the ones which remained after the first elimination procedure, we choose those, if there are any, which contain a point $\boldsymbol{x}_k$ belonging to E such that $J_k \subseteq B[\boldsymbol{x}_k, \delta(\boldsymbol{x}_k)]$; once this choice has been made, we eliminate all the smaller rectangles contained in them; and so on. □

Remark Notice that if, in the assumptions of the covering lemma, it happens that E is contained in an open set which is itself contained in I, then all the rectangles J_k can be chosen so to be all contained in that open set.

We can now prove the following characterization of the bounded measurable sets.[1]

Proposition 2.21
Let E be a bounded set, contained in a rectangle I. The three following propositions are equivalent:

(*i*) *the set E is measurable;*

(*ii*) *for every $\varepsilon > 0$ there are two finite or countable families (J_k) and (J'_k), each made of [non-overlapping] rectangles contained in I, such that*

$$E \subseteq \bigcup_k J_k\,,\quad I \setminus E \subseteq \bigcup_k J'_k \quad \text{and} \quad \mu\left(\left(\bigcup_k J_k\right) \cap \left(\bigcup_k J'_k\right)\right) \leq \varepsilon\,;$$

(*iii*) *there are two sequences $(E_n)_{n\geq 1}$ and $(E'_n)_{n\geq 1}$ of bounded and measurable subsets such that*

$$E'_n \subseteq E \subseteq E_n\,,\qquad \lim_{n\to\infty} (\mu(E_n) - \mu(E'_n)) = 0\,.$$

In that case, we have:

$$\mu(E) = \lim_{n\to\infty} \mu(E_n) = \lim_{n\to\infty} \mu(E'_n)\,.$$

Proof
Let us first prove that (i) implies (ii). Assume that E be measurable, and fix $\varepsilon > 0$. By the Saks–Henstock's theorem, there is a gauge δ on I such that, for every δ-fine sub-P-partition $\Pi = \{(\boldsymbol{x}_j, K_j) : j = 1, \dots, m\}$ of I,

$$\sum_{j=1}^{m} \left| \chi_E(\boldsymbol{x}_j)\mu(K_j) - \int_{K_j} \chi_E \right| \leq \frac{\varepsilon}{2}\,.$$

By the covering lemma, there is a family of non-overlapping rectangles J_k, contained in I, whose union covers E and in each J_k there is a point $\boldsymbol{x}_k$ belonging to E such that $J_k \subseteq B[\boldsymbol{x}_k, \delta(\boldsymbol{x}_k)]$. Let us fix a positive integer N and consider only $(\boldsymbol{x}_1, J_1), \dots, (\boldsymbol{x}_N, J_N)$. They

[1] In the following statements, the words in squared brackets may be omitted.

make up a δ-fine sub-P-partition of I. From the above inequality we then deduce that

$$\sum_{k=1}^{N} \left| \mu(J_k) - \int_{J_k} \chi_E \right| \leq \frac{\varepsilon}{2},$$

whence

$$\sum_{k=1}^{N} \mu(J_k) \leq \sum_{k=1}^{N} \int_{J_k} \chi_E + \frac{\varepsilon}{2} \leq \int_I \chi_E + \frac{\varepsilon}{2} = \mu(E) + \frac{\varepsilon}{2}.$$

Since this holds for every positive integer N, we have thus constructed a family (J_k) of non-overlapping rectangles, such that

$$E \subseteq \bigcup_k J_k, \quad \sum_k \mu(J_k) \leq \mu(E) + \frac{\varepsilon}{2}.$$

Consider now $I \setminus E$, which is measurable, as well. We can repeat the same procedure as above replacing E by $I \setminus E$, thus finding a family (J_k') of non-overlapping rectangles, contained in I, such that

$$I \setminus E \subseteq \bigcup_k J_k', \quad \sum_k \mu(J_k') \leq \mu(I \setminus E) + \frac{\varepsilon}{2}.$$

Consequently,

$$I \setminus \left(\bigcup_k J_k' \right) \subseteq E \subseteq \bigcup_k J_k,$$

and hence

$$\begin{aligned}
\mu\left(\left(\bigcup_k J_k \right) \cap \left(\bigcup_k J_k' \right) \right) &= \mu\left(\left(\bigcup_k J_k \right) \setminus \left(I \setminus \left(\bigcup_k J_k' \right) \right) \right) \\
&= \mu\left(\bigcup_k J_k \right) - \mu\left(I \setminus \left(\bigcup_k J_k' \right) \right) \\
&= \mu\left(\bigcup_k J_k \right) - \mu(I) + \mu\left(\bigcup_k J_k' \right) \\
&\leq \left(\mu(E) + \frac{\varepsilon}{2} \right) - \mu(I) + \left(\mu(I \setminus E) + \frac{\varepsilon}{2} \right) \\
&= \varepsilon,
\end{aligned}$$

and the implication is thus proved.

Taking $\varepsilon = \frac{1}{n}$, it is easy to see that (ii) implies (iii). Let us prove now that (iii) implies (i). Consider the measurable sets

$$\widetilde{E}' = \bigcup_{n\geq 1} E'_n \,, \qquad \widetilde{E} = \bigcap_{n\geq 1} E_n \,,$$

for which it has to be

$$\widetilde{E}' \subseteq E \subseteq \widetilde{E} \,, \qquad \mu(\widetilde{E}') = \mu(\widetilde{E}) \,.$$

Equivalently, we have

$$\chi_{\widetilde{E}'} \leq \chi_E \leq \chi_{\widetilde{E}} \,, \qquad \int_I (\chi_{\widetilde{E}} - \chi_{\widetilde{E}'}) = 0 \,,$$

so that $\chi_{\widetilde{E}'} = \chi_E = \chi_{\widetilde{E}}$ almost everywhere. Then, E is measurable and $\mu(E) = \mu(\widetilde{E}') = \mu(\widetilde{E})$. Moreover,

$$0 \leq \lim_n [\mu(E) - \mu(E'_n)] \leq \lim_n [\mu(E_n) - \mu(E'_n)] = 0 \,,$$

hence $\mu(E) = \lim_n \mu(E'_n)$. Analogously it is seen that $\mu(E) = \lim_n \mu(E_n)$, and the proof is thus completed. □

Proposition 2.22

Let E be a bounded set. Then, E is negligible if and only if for every $\varepsilon > 0$ there is a finite or countable family (J_k) of [non-overlapping] rectangles such that

$$E \subseteq \bigcup_k J_k \,, \qquad \sum_k \mu(J_k) \leq \varepsilon \,.$$

Proof

The necessary condition is proved in the first part of the proof of the previous proposition.

Let us prove the sufficiency. Once fixed $\varepsilon > 0$, assume there exists a family (J_k) with the given properties. Let I be a rectangle containing the set E. On the other hand, consider a family (J'_k) whose elements all coincide with I. The condition (ii) of the previous proposition is then satisfied, so that E is indeed measurable. Then,

$$\mu(E) \leq \mu\left(\bigcup_k J_k\right) \leq \sum_k \mu(J_k) \leq \varepsilon \,;$$

being ε arbitrary, it has to be $\mu(E) = 0$. □

Remark Observe that if E is contained in an open set which is itself contained in a rectangle I, all the rectangles J_k can be chosen so to be all contained in that open set.

As a consequence of the previous proposition, it is not difficult to prove the following.

Corollary 2.23
If I_{N-1} is a rectangle in $\mathbb{R}^{N-1}$ and T is a negligible subset of $\mathbb{R}$, then $I_{N-1} \times T$ is negligible in $\mathbb{R}^N$.

Proof
Fix $\varepsilon > 0$ and, according to Proposition 2.22, let (J_k) be a finite or countable family of intervals in $\mathbb{R}$ such that

$$T \subseteq \bigcup_k J_k\,, \qquad \sum_k \mu(J_k) \leq \frac{\varepsilon}{\mu(I_{N-1})}\,.$$

Defining the rectangles $\tilde{J}_k = I_{N-1} \times J_k$, we have that

$$I_{N-1} \times T \subseteq \bigcup_k \tilde{J}_k\,, \qquad \sum_k \mu(\tilde{J}_k) = \mu(I_{N-1}) \sum_k \mu(J_k) \leq \mu(I_{N-1}) \frac{\varepsilon}{\mu(I_{N-1})} = \varepsilon\,,$$

and Proposition 2.22 applies again. □

2.7 Continuous Functions and L-Integrable Functions

We begin this section by showing that the continuous functions are L-integrable on compact sets.

Theorem 2.24
Let $E \subseteq \mathbb{R}^N$ be a compact set and $f : E \to \mathbb{R}$ be a continuous function. Then, f is L-integrable on E.

Proof
We consider for simplicity the case $N = 2$. Being f continuous on a compact set, there is a constant $C > 0$ such that

$$|f(x)| \leq C\,,$$

for every $\boldsymbol{x} \in E$. Let I be a rectangle containing E. First we divide I into four rectangles, by tracing the segments joining the mid points of its edges; we denote by $U_{1,1}, U_{1,2}, U_{1,3}, U_{1,4}$ these subrectangles. We now divide again each of these rectangles in the same way, thus obtaining sixteen smaller subrectangles, which we denote by $U_{2,1}, U_{2,2}, \ldots, U_{2,16}$. Proceeding in this way, for every k we will have a subdivision of the rectangle I in 2^{2k} small rectangles $U_{k,j}$, with $j = 1, \ldots, 2^{2k}$. Whenever E has non-empty intersection with $\mathring{U}_{k,j}$, we choose and fix a point $\boldsymbol{x}_{k,j} \in E \cap \mathring{U}_{k,j}$. Define now the function f_k in the following way:

- if $E \cap \mathring{U}_{k,j}$ is non-empty, f_k is constant on $\mathring{U}_{k,j}$ with value $f(\boldsymbol{x}_{k,j})$;
- if $E \cap \mathring{U}_{k,j}$ is empty, f_k is constant on $\mathring{U}_{k,j}$ with value 0.

The functions f_k are thus defined almost everywhere on I, not being defined only on the points of the grid made up by the above constructed segments, which form a countable family of negligible sets. The functions f_k are integrable on each subrectangle $U_{k,j}$, being constant in its interior. By the property of additivity on subrectangles, these functions are therefore integrable on I. Moreover,

$$|f_k(\boldsymbol{x})| \leq C\,,$$

for almost every $\boldsymbol{x} \in I$ and every $k \geq 1$. Let us see now that f_k converges pointwise almost everywhere to f_E. Indeed, taking a point $\boldsymbol{x} \in I$ not belonging to the grid, for every k there is a $j = j(k)$ for which $\boldsymbol{x} \in \mathring{U}_{k,j(k)}$; we have two possibilities:

a) $\boldsymbol{x} \notin E$; in this case, being E closed, we have that, for k sufficiently large, $\mathring{U}_{k,j(k)}$ (whose dimensions tend to zero as $k \to \infty$) will have empty intersection with E, and then $f_k(\boldsymbol{x}) = 0 = f_E(\boldsymbol{x})$.
b) $\boldsymbol{x} \in E$; in this case, if $k \to +\infty$, we have that $\boldsymbol{x}_{k,j(k)} \to \boldsymbol{x}$ (again using the fact that $\mathring{U}_{k,j(k)}$ has dimensions tending to zero). By the continuity of f, we have that

$$f_k(\boldsymbol{x}) = f(\boldsymbol{x}_{k,j(k)}) \to f(\boldsymbol{x}) = f_E(\boldsymbol{x})\,.$$

The Dominated Convergence Theorem then yields the conclusion. □

We now see that the L-integrability is conserved on measurable subsets.

Theorem 2.25
Let $f : E \to \mathbb{R}$ be a L-integrable function on a bounded set E. Then, f is L-integrable on every measurable subset of E.

Proof
Assume first f to have non-negative values. Let S be a measurable subset of E, and define on E the functions $f_k = \min\{f, k\chi_S\}$. They form an increasing sequence of L-integrable

functions, since such are both f and $k\chi_S$, which converges pointwise to f_S. Moreover, it is

$$\int_E f_k \leq \int_E f\,,$$

for every k. The Monotone Convergence Theorem then guarantees that f is integrable on S in this case. In the general case, f being L-integrable, both f^+ and f^- are L-integrable on E. Hence, by the above, they are both L-integrable on S, and then such is f, too. □

Let us prove now the property of **complete additivity of the integral for L-integrable functions**. We will say that two bounded measurable subsets are **non-overlapping** if their intersection is a negligible set.

Theorem 2.26
Let (E_k) be a finite or countable family of measurable non-overlapping sets whose union is a bounded set E. Then f is L-integrable on E if and only if the two following conditions hold:
(a) f is L-integrable on each E_k ;
(b) $\sum_k \int_{E_k} |f(\boldsymbol{x})|\,d\boldsymbol{x} < +\infty$.

In that case, we have:

$$\int_E f = \sum_k \int_{E_k} f\,.$$

Proof
Observe that

$$f(\boldsymbol{x}) = \sum_k f_{E_k}(\boldsymbol{x})\,, \qquad |f(\boldsymbol{x})| = \sum_k |f_{E_k}(\boldsymbol{x})|\,,$$

for almost every $\boldsymbol{x} \in E$. If f is L-integrable on E, from the preceding theorem, (a) follows. Moreover, it is obvious that (b) holds whenever the sets E_k are in a finite number. If instead they are infinite, for any fixed n, we have

$$\sum_{k=1}^{n} \int_{E_k} |f(\boldsymbol{x})|\,d\boldsymbol{x} = \sum_{k=1}^{n} \int_E |f_{E_k}(\boldsymbol{x})|\,d\boldsymbol{x} \leq \int_E |f(\boldsymbol{x})|\,d\boldsymbol{x}\,,$$

and (b) follows.

Assume now that (a) and (b) hold. If the sets E_k are in a finite number, it is sufficient to integrate on E both terms in the equation $f = \sum_k f_{E_k}$. If instead they are infinite, assume first that f has non-negative values. In this case, the corollary following the Monotone

Convergence Theorem, when applied to the series $\sum_k f_{E_k}$, yields the conclusion. In the general case, it is sufficient to consider, as usual, the positive and the negative parts of f.

□

Exercises

1. Compute the area of the set

$$\{(x, y) \in \mathbb{R}^2 : 0 \le x \le 1 \,,\; 0 \le y \le x\}\,.$$

2. Compute the following integral:

$$\int_{[0,1]\times[0,1]} x^2 \, dx \, dy\,.$$

3. Prove that the function $f :]0, 1] \times [0, 1] \to \mathbb{R}$, defined as

$$f(x, y) = \sqrt{k} \quad \text{if} \quad (x, y) \in \left]\frac{1}{k+1}, \frac{1}{k}\right] \times [0, 1]\,,$$

is integrable, and give an estimate of its integral.

2.8 Limits and Derivatives Under the Integration Sign

Let X be a metric space, Y a bounded subset of $\mathbb{R}^N$, and consider a function $f : X \times Y \to \mathbb{R}$. (For simplicity, we may think of X and Y as subsets of $\mathbb{R}$.) The first question we want to face is the following: when does the formula

$$\lim_{x \to x_0} \left(\int_Y f(x, y)\, dy \right) = \int_Y \left(\lim_{x \to x_0} f(x, y) \right) dy$$

hold? The following is a generalization of the Dominated Convergence Theorem.

Theorem 2.27

Let x_0 be an accumulation point of X, and assume that:

(i) for every $x \in X \setminus \{x_0\}$, the function $f(x, \cdot)$ is integrable on Y, so that we can define the function

$$F(x) = \int_Y f(x, y)\, dy\,;$$

(ii) for almost every $y \in Y$ the limit $\lim_{x \to x_0} f(x, y)$ exists and is finite, so that we can define almost everywhere the function

$$\eta(y) = \lim_{x \to x_0} f(x, y)\,;$$

(Continued)

Theorem 2.27 (continued)

(*iii*) *there are two integrable functions* $g, h : Y \to \mathbb{R}$ *such that*

$$g(y) \le f(x, y) \le h(y),$$

for every $x \in X \setminus \{x_0\}$ *and almost every* $y \in Y$.

Then, η *is integrable on* Y*, and we have:*

$$\lim_{x \to x_0} F(x) = \int_Y \eta(y)\, dy .$$

Proof

Let us take a sequence $(x_k)_k$ in $X \setminus \{x_0\}$ which tends to x_0. Define, for every k, the functions $f_k : Y \to \mathbb{R}$ such that $f_k(y) = f(x_k, y)$. By the assumptions (i), (ii) and (iii) we can apply the Dominated Convergence Theorem, so that

$$\lim_k F(x_k) = \lim_k \left(\int_Y f_k(y)\, dy \right) = \int_Y \left(\lim_k f_k(y) \right) dy = \int_Y \eta(y)\, dy .$$

The conclusion then follows from the characterization of the limit by the use of sequences. □

We have the following consequence.

Corollary 2.28

If X *is a subset of* $\mathbb{R}^M$, $Y \subseteq \mathbb{R}^N$ *is compact, and* $f : X \times Y \to \mathbb{R}$ *is continuous, then the function* $F : X \to \mathbb{R}$*, defined by*

$$F(x) = \int_Y f(x, y)\, dy ,$$

is continuous.

Proof

The function $F(x)$ is well defined, being $f(x, \cdot)$ continuous on the compact set Y. Let us fix $x_0 \in X$ and prove that F is continuous at x_0. By the continuity of f,

$$\eta(y) = \lim_{x \to x_0} f(x, y) = f(x_0, y) .$$

Moreover, once taken a compact neighborhood U of x_0, there is a constant $C > 0$ such that $|f(x, y)| \leq C$, for every $(x, y) \in U \times Y$. Theorem 2.27 can then be applied, and we have:

$$\lim_{x \to x_0} F(x) = \int_Y f(x_0, y)\, dy = F(x_0),$$

thus proving that F is continuous at x_0. □

Let now X be a subset of $\mathbb{R}$. The second question we want to face is the following: when does the formula

$$\frac{d}{dx}\left(\int_Y f(x, y)\, dy\right) = \int_Y \left(\frac{\partial f}{\partial x}(x, y)\right) dy$$

hold? The following result is often quoted as **the Leibniz rule**.

Theorem 2.29
Let X be an interval in $\mathbb{R}$ containing x_0, $Y \subseteq \mathbb{R}^N$ be bounded, and assume that:
(i) for every $x \in X$, the function $f(x, \cdot)$ is integrable on Y, so that we can define the function

$$F(x) = \int_Y f(x, y)\, dy ;$$

(ii) for every $x \in X$ and almost every $y \in Y$, the partial derivative $\frac{\partial f}{\partial x}(x, y)$ exists;
(iii) there are two integrable functions $g, h : Y \to \mathbb{R}$ such that

$$g(y) \leq \frac{\partial f}{\partial x}(x, y) \leq h(y),$$

for every $x \in X$ and almost every $y \in Y$.

Then, the function $\frac{\partial f}{\partial x}(x, \cdot)$, defined almost everywhere on Y, is integrable there, the derivative of F at x_0 exists, and we have:

$$F'(x_0) = \int_Y \left(\frac{\partial f}{\partial x}(x_0, y)\right) dy .$$

Proof
We define, for $x \in X$ different from x_0, the function

$$\psi(x, y) = \frac{f(x, y) - f(x_0, y)}{x - x_0} .$$

For every $x \in X \setminus \{x_0\}$, the function $\psi(x,\cdot)$ is integrable on Y. Moreover, for almost every $y \in Y$, it is

$$\lim_{x\to x_0} \psi(x,y) = \frac{\partial f}{\partial x}(x_0,y)\,.$$

By the Lagrange Mean Value Theorem, for (x,y) as above there is a $\xi \in X$ between x_0 and x such that

$$\psi(x,y) = \frac{\partial f}{\partial x}(\xi,y)\,.$$

By assumption (iii), we then have

$$g(y) \le \psi(x,y) \le h(y)\,,$$

for every $x \in X \setminus \{x_0\}$ and almost every $y \in Y$. We are then in the hypotheses of the previous theorem, and we can conclude that the function $\frac{\partial f}{\partial x}(x_0,\cdot)$, defined almost everywhere on Y, is integrable there, and

$$\lim_{x\to x_0}\left(\int_Y \psi(x,y)\,dy\right) = \int_Y \left(\frac{\partial f}{\partial x}(x_0,y)\right) dy\,.$$

On the other hand,

$$\begin{aligned}\lim_{x\to x_0}\left(\int_Y \psi(x,y)\,dy\right) &= \lim_{x\to x_0}\left(\int_Y \frac{f(x,y)-f(x_0,y)}{x-x_0}\,dy\right)\\ &= \lim_{x\to x_0}\frac{1}{x-x_0}\left(\int_Y f(x,y)\,dy - \int_Y f(x_0,y)\,dy\right)\\ &= \lim_{x\to x_0}\frac{F(x)-F(x_0)}{x-x_0}\,,\end{aligned}$$

so that F is differentiable at x_0 and the conclusion holds true. □

Corollary 2.30

If X is an interval in $\mathbb{R}$, Y is a compact subset of $\mathbb{R}^N$, and the function $f : X \times Y \to \mathbb{R}$ is continuous and has a continuous partial derivative $\frac{\partial f}{\partial x} : X \times Y \to \mathbb{R}$, then the function $F : X \to \mathbb{R}$, defined by

$$F(x) = \int_Y f(x,y)\,dy\,,$$

is differentiable with a continuous derivative.

Proof

The function $F(x)$ is well defined, being $f(x,\cdot)$ continuous on the compact set Y. Taking a point $x_0 \in X$ and a nontrivial compact interval $U \subseteq X$ containing it, there is a constant $C > 0$ such that $|\frac{\partial f}{\partial x}(x,y)| \leq C$, for every $(x,y) \in U \times Y$. By Theorem 2.27, F is differentiable at x_0. The same argument holds replacing x_0 by any $x \in X$, and

$$F'(x) = \int_Y \left(\frac{\partial f}{\partial x}(x,y)\right) dy\,.$$

The continuity of $F' : X \to \mathbb{R}$ now follows from Corollary 2.28. □

Example Consider, for $x \geq 0$, the function

$$f(x,y) = \frac{e^{-x^2(y^2+1)}}{y^2+1}\,.$$

We want to see if the corresponding function $F(x) = \int_0^1 f(x,y)\,dy$ is differentiable and, in this case, to find its derivative. We have that

$$\frac{\partial f}{\partial x}(x,y) = -2xe^{-x^2(y^2+1)}\,,$$

which, for $y \in [0,1]$ and $x \geq 0$, is such that

$$-\sqrt{\frac{2}{e}} \leq -2xe^{-x^2} \leq -2xe^{-x^2(y^2+1)} \leq 0\,.$$

We can then apply the Leibniz rule, so that

$$F'(x) = -2x \int_0^1 e^{-x^2(y^2+1)}\,dy\,.$$

Now that we have found the derivative of $F(x)$, let us make a digression, so to present a nice formula. By the change of variable $t = xy$, one has

$$-2x \int_0^1 e^{-x^2(y^2+1)}\,dy = -2e^{-x^2} \int_0^x e^{-t^2}\,dt = -\frac{d}{dx}\left(\int_0^x e^{-t^2}\,dt\right)^2.$$

Taking into account that $F(0) = \pi/4$, we can write

$$F(x) = \frac{\pi}{4} - \left(\int_0^x e^{-t^2}\,dt\right)^2.$$

We would like now to pass to the limit for $x \to +\infty$. Since, for $x \geq 0$, one has

$$0 \leq \frac{e^{-x^2(y^2+1)}}{y^2+1} \leq 1\,,$$

we can pass to the limit under the sign of integration, thus obtaining

$$\lim_{x\to+\infty}\int_0^1 \frac{e^{-x^2(y^2+1)}}{y^2+1}\,dy = \int_0^1 \left(\lim_{x\to+\infty}\frac{e^{-x^2(y^2+1)}}{y^2+1}\right)dy = 0\,.$$

Hence,

$$\left(\int_0^{+\infty} e^{-t^2}\,dt\right)^2 = \frac{\pi}{4}$$

and, by symmetry,

$$\int_{-\infty}^{+\infty} e^{-t^2}\,dt = \sqrt{\pi}\,,$$

which is a very useful formula in various applications.

2.9 The Reduction Formula

Before stating the main theorem, it is useful to first prove a preliminary result.

Proposition 2.31
Let $f : I \to \mathbb{R}$ be an integrable function on the rectangle $I = [a_1, b_1] \times [a_2, b_2]$. Then, for almost every $x \in [a_1, b_1]$, the function $f(x, \cdot)$ is integrable on $[a_2, b_2]$.

Proof
Let $T \subseteq [a_1, b_1]$ be the set of those $x \in [a_1, b_1]$ for which $f(x, \cdot)$ is not integrable on $[a_2, b_2]$. Let us prove that T is a negligible set. For each $x \in T$, the Cauchy condition does not hold. Hence, if we define the sets

$$T_n = \left\{ x : \begin{array}{l} \text{for every gauge } \delta_2 \text{ on } [a_2, b_2] \text{ there are two} \\ \delta_2\text{-fine P-partitions } \Pi_2 \text{ and } \widetilde{\Pi}_2 \text{ of } [a_2, b_2] \text{ such that} \\ S([a_2, b_2], f(x, \cdot), \Pi_2) - S([a_2, b_2], f(x, \cdot), \widetilde{\Pi}_2) > \frac{1}{n} \end{array} \right\},$$

we have that each $x \in T$ belongs to T_n, if n is sufficiently large. So, T is the union of all T_n, and if we prove that any T_n is negligible, by the properties of the measure we will also have that T is such. In order to do so, let us consider a certain T_n and fix $\varepsilon > 0$. Being f integrable on I, there is a gauge δ on I such that, taken two δ-fine P-partitions Π and $\widetilde{\Pi}$ of I, it is

$$|S(I, f, \Pi) - S(I, f, \widetilde{\Pi})| \le \frac{\varepsilon}{n}\,.$$

The gauge δ on I determines, for every $x \in [a_1, b_1]$, a gauge $\delta(x, \cdot)$ on $[a_2, b_2]$. We now associate to each $x \in [a_1, b_1]$ two $\delta(x, \cdot)$-fine P-partitions Π_2^x and $\widetilde{\Pi}_2^x$ of $[a_2, b_2]$ in the following way:

– if $x \in T_n$, we can choose Π_2^x and $\widetilde{\Pi}_2^x$ such that

$$S([a_2, b_2], f(x, \cdot), \Pi_2^x) - S([a_2, b_2], f(x, \cdot), \widetilde{\Pi}_2^x) > \frac{1}{n};$$

– if instead $x \notin T_n$, we take Π_2^x and $\widetilde{\Pi}_2^x$ equal to each other.

Let us write the two P-partitions Π_2^x and $\widetilde{\Pi}_2^x$ thus determined:

$$\Pi_2^x = \{(y_j^x, K_j^x) : j = 1, \dots, m^x\}, \quad \widetilde{\Pi}_2^x = \{(\tilde{y}_j^x, \widetilde{K}_j^x) : j = 1, \dots, \widetilde{m}^x\}.$$

We define a gauge δ_1 on $[a_1, b_1]$, setting

$$\delta_1(x) = \min\{\delta(x, y_1^x), \dots, \delta(x, y_{m^x}^x), \delta(x, \tilde{y}_1^x), \dots, \delta(x, \tilde{y}_{\widetilde{m}^x}^x)\}.$$

Let now $\Pi_1 = \{(x_i, J_i) : i = 1, \dots, k\}$ be a δ_1-fine P-partition of $[a_1, b_1]$. We want to prove that $S([a_1, b_1], \chi_{T_n}, \Pi_1) \leq \varepsilon$, i.e.

$$\sum_{\{i \,:\, x_i \in T_n\}} \mu(J_i) \leq \varepsilon.$$

To this aim, define the following two P-partitions of I which make use of the elements of Π_1:

$$\Pi = \{((x_i, y_j^{x_i}), J_i \times K_j^{x_i}) : i = 1, \dots, k, \; j = 1, \dots, m^{x_i}\},$$

$$\widetilde{\Pi} = \{((x_i, \tilde{y}_j^{x_i}), J_i \times \widetilde{K}_j^{x_i}) : i = 1, \dots, k, \; j = 1, \dots, \widetilde{m}^{x_i}\}.$$

They are δ-fine, and hence

$$|S(I, f, \Pi) - S(I, f, \widetilde{\Pi})| \leq \frac{\varepsilon}{n}.$$

On the other hand, we have

$$\begin{aligned}
|S(I, f, \Pi) &- S(I, f, \widetilde{\Pi})| = \\
&= \left| \sum_{i=1}^{k} \sum_{j=1}^{m^{x_i}} f(x_i, y_j^{x_i}) \mu(J_i \times K_j^{x_i}) - \sum_{i=1}^{k} \sum_{j=1}^{\widetilde{m}^{x_i}} f(x_i, \tilde{y}_j^{x_i}) \mu(J_i \times \widetilde{K}_j^{x_i}) \right| \\
&= \left| \sum_{i=1}^{k} \mu(J_i) \left[\sum_{j=1}^{m^{x_i}} f(x_i, y_j^{x_i}) \mu(K_j^{x_i}) - \sum_{j=1}^{\widetilde{m}^{x_i}} f(x_i, \tilde{y}_j^{x_i}) \mu(\widetilde{K}_j^{x_i}) \right] \right|
\end{aligned}$$

$$
= \left| \sum_{i=1}^{k} \mu(J_i)[S([a_2, b_2], f(x_i, \cdot), \Pi_2^{x_i}) - S([a_2, b_2], f(x_i, \cdot), \widetilde{\Pi}_2^{x_i})] \right|
$$

$$
= \sum_{\{i \,:\, x_i \in T_n\}} \mu(J_i)[S([a_2, b_2], f(x_i, \cdot), \Pi_2^{x_i}) - S([a_2, b_2], f(x_i, \cdot), \widetilde{\Pi}_2^{x_i})] .
$$

Recalling that, if $x_i \in T_n$,

$$
S([a_2, b_2], f(x_i, \cdot), \Pi_2^{x_i}) - S([a_2, b_2], f(x_i, \cdot), \widetilde{\Pi}_2^{x_i}) > \frac{1}{n} ,
$$

we conclude that

$$
\frac{\varepsilon}{n} \geq |S(I, f, \Pi) - S(I, f, \widetilde{\Pi})| > \frac{1}{n} \sum_{\{i \,:\, x_i \in T_n\}} \mu(J_i) ,
$$

and hence $S([a_1, b_1], \chi_{T_n}, \Pi_1) \leq \varepsilon$, which is what we wanted to prove. All this shows that the sets T_n are negligible, and therefore T is negligible, too. □

The following **Reduction Theorem**, due to G. Fubini, permits to compute the integral of an integrable function of two variables by performing two integrations of functions of one variable.

Theorem 2.32

Let $f : I \to \mathbb{R}$ be an integrable function on the rectangle $I = [a_1, b_1] \times [a_2, b_2]$. Then:

(i) *for almost every $x \in [a_1, b_1]$, the function $f(x, \cdot)$ is integrable on $[a_2, b_2]$;*

(ii) *the function $\int_{a_2}^{b_2} f(\cdot, y)\, dy$, defined almost everywhere on $[a_1, b_1]$, is integrable there;*

(iii) *we have*

$$
\int_I f = \int_{a_1}^{b_1} \left(\int_{a_2}^{b_2} f(x, y)\, dy \right) dx .
$$

Proof

We have already proved (i) in the preliminary proposition. Let us now prove (ii) and (iii). Let T be the negligible subset of $[a_1, b_1]$ such that, for $x \in T$, the function $f(x, \cdot)$ is not integrable on $[a_2, b_2]$. Being $T \times [a_2, b_2]$ negligible in I, we can modify on that set the function f without changing the integrability properties. (We can choose, for example, $f = 0$ on that set.) In this way, we can assume without loss of generality that T be empty. Let us define

$$
F(x) = \int_{a_2}^{b_2} f(x, y)\, dy .
$$

We want to prove that F is integrable on $[a_1, b_1]$ and that

$$\int_{a_1}^{b_1} F = \int_I f .$$

Let $\varepsilon > 0$ be fixed. Because of the integrability of f on I, there is a gauge δ on I such that, for every δ-fine P-partition Π of I,

$$\left| S(I, f, \Pi) - \int_I f \right| \leq \frac{\varepsilon}{2} .$$

We now associate to each $x \in [a_1, b_1]$ a $\delta(x, \cdot)$-fine P-partition Π_2^x of $[a_2, b_2]$ such that

$$|S([a_2, b_2], f(x, \cdot), \Pi_2^x) - F(x)| \leq \frac{\varepsilon}{2(b_1 - a_1)}$$

(this is possible since $f(x, \cdot)$ is integrable on $[a_2, b_2]$ with integral $F(x)$). Let us write the P-partitions thus determined:

$$\Pi_2^x = \{(y_j^x, K_j^x) : j = 1, \dots, m^x\} .$$

We define a gauge δ_1 on $[a_1, b_1]$, by setting

$$\delta_1(x) = \min\{\delta(x, y_1^x), \dots, \delta(x, y_{m^x}^x)\} .$$

Consider now a δ_1-fine P-partition of $[a_1, b_1]$:

$$\Pi_1 = \{(x_i, J_i) : i = 1, \dots, n\} ,$$

and construct the following δ-fine P-partition of I, making use of the elements of Π_1 :

$$\Pi = \{((x_i, y_j^{x_i}), J_i \times K_j^{x_i}) : i = 1, \dots, n\ ,\ \ j = 1, \dots, m^{x_i}\} .$$

We have the following inequalities:

$$\begin{aligned} \left| S([a_1, b_1], F, \Pi_1) - \int_I f \right| &\leq \\ &\leq |S([a_1, b_1], F, \Pi_1) - S(I, f, \Pi)| + \left| S(I, f, \Pi) - \int_I f \right| \\ &\leq \left| \sum_{i=1}^n F(x_i)\mu(J_i) - \sum_{i=1}^n \sum_{j=1}^{m^{x_i}} f(x_i, y_j^{x_i})\mu(J_i \times K_j^{x_i}) \right| + \frac{\varepsilon}{2} \end{aligned}$$

$$\leq \sum_{i=1}^{n} \left| F(x_i) - \sum_{j=1}^{m^{x_i}} f(x_i, y_j^{x_i}) \mu(K_j^{x_i}) \right| \mu(J_i) + \frac{\varepsilon}{2}$$

$$\leq \sum_{i=1}^{n} \frac{\varepsilon}{2(b_1 - a_1)} \mu(J_i) + \frac{\varepsilon}{2} = \varepsilon \,.$$

This proves that F is integrable on $[a_1, b_1]$ and

$$\int_{a_1}^{b_1} F = \int_I f \,.$$

The proof is thus completed. □

Example Consider the function $f(x, y) = x^2 \sin y$ on the rectangle $I = [-1, 1] \times [0, \pi]$. Being f continuous on a compact set, it is integrable there, so that:

$$\int_I f = \int_{-1}^{1} \left(\int_0^{\pi} x^2 \sin y \, dy \right) dx$$

$$= \int_{-1}^{1} x^2 [-\cos y]_0^{\pi} \, dx = 2 \int_{-1}^{1} x^2 \, dx = 2 \left[\frac{x^3}{3} \right]_{-1}^{1} = \frac{4}{3} \,.$$

Clearly, the following version of the Reduction Theorem holds, which is symmetric with respect to the preceding one.

Theorem 2.33

Let $f : I \to \mathbb{R}$ be an integrable function on the rectangle $I = [a_1, b_1] \times [a_2, b_2]$. Then:

(i) *for almost every $y \in [a_2, b_2]$, the function $f(\cdot, y)$ is integrable on $[a_1, b_1]$;*

(ii) *the function $\int_{a_1}^{b_1} f(x, \cdot) \, dx$, defined almost everywhere on $[a_2, b_2]$, is integrable there;*

(iii) *we have:*

$$\int_I f = \int_{a_2}^{b_2} \left(\int_{a_1}^{b_1} f(x, y) \, dx \right) dy \,.$$

As an immediate consequence, we have that, if f is integrable on $I = [a_1, b_1] \times [a_2, b_2]$, then

$$\int_{a_1}^{b_1} \left(\int_{a_2}^{b_2} f(x, y) \, dy \right) dx = \int_{a_2}^{b_2} \left(\int_{a_1}^{b_1} f(x, y) \, dx \right) dy \,.$$

Therefore, if the above equality does not hold, then the function f is not integrable on I.

Examples Consider the function

$$f(x,y)=\begin{cases}\dfrac{x^2-y^2}{(x^2+y^2)^2} & \text{if } (x,y)\neq(0,0)\,,\\ 0 & \text{if } (x,y)=(0,0)\,,\end{cases}$$

on the rectangle $I=[0,1]\times[0,1]$. If $x\neq 0$, it is

$$\int_0^1 \frac{x^2-y^2}{(x^2+y^2)^2}\,dy=\left[\frac{y}{x^2+y^2}\right]_{y=0}^{y=1}=\frac{1}{x^2+1}\,,$$

so that

$$\int_0^1\left(\int_0^1 \frac{x^2-y^2}{(x^2+y^2)^2}\,dy\right)dx=\int_0^1\frac{1}{x^2+1}\,dx=[\arctan x]_0^1=\frac{\pi}{4}\,.$$

Analogously, we see that

$$\int_0^1\left(\int_0^1 \frac{x^2-y^2}{(x^2+y^2)^2}\,dx\right)dy=-\frac{\pi}{4}\,,$$

and we thus conclude that f is not integrable on I.

As a further example, consider the function

$$f(x,y)=\begin{cases}\dfrac{xy}{(x^2+y^2)^2} & \text{if } (x,y)\neq(0,0)\,,\\ 0 & \text{if } (x,y)=(0,0)\,,\end{cases}$$

on the rectangle $I=[-1,1]\times[-1,1]$. In this case, if $x\neq 0$, we have

$$\int_{-1}^1 \frac{xy}{(x^2+y^2)^2}\,dy=\left[\frac{-x}{2(x^2+y^2)}\right]_{y=-1}^{y=1}=0\,,$$

so that

$$\int_{-1}^1\left(\int_{-1}^1 \frac{xy}{(x^2+y^2)^2}\,dy\right)dx=0\,.$$

Analogously, we see that

$$\int_{-1}^1\left(\int_{-1}^1 \frac{xy}{(x^2+y^2)^2}\,dx\right)dy=0\,.$$

Nevertheless, we are not allowed to conclude that f is integrable on I. Truly, it is not at all. Indeed, if f were integrable, it should be such on every subrectangle, and in particular on

$[0, 1] \times [0, 1]$. But, if $x \neq 0$, we have

$$\int_0^1 \frac{xy}{(x^2+y^2)^2}\,dy = \left[\frac{-x}{2(x^2+y^2)}\right]_{y=0}^{y=1} = \frac{1}{2x(x^2+1)}\,,$$

which is not integrable with respect to x on $[0, 1]$.

When the function f is defined on a bounded subset E of $\mathbb{R}^2$, it is possible to state the Reduction Theorem for the function f_E. Let $I = [a_1, b_1] \times [a_2, b_2]$ be a rectangle containing E. Let us define the **sections** of E :

$$E_x = \{y \in [a_2, b_2] : (x, y) \in E\}\,, \qquad E_y = \{x \in [a_1, b_1] : (x, y) \in E\}\,,$$

and the **projections** of E :

$$P_1 E = \{x \in [a_1, b_1] : E_x \neq \emptyset\}\,, \qquad P_2 E = \{y \in [a_2, b_2] : E_y \neq \emptyset\}\,.$$

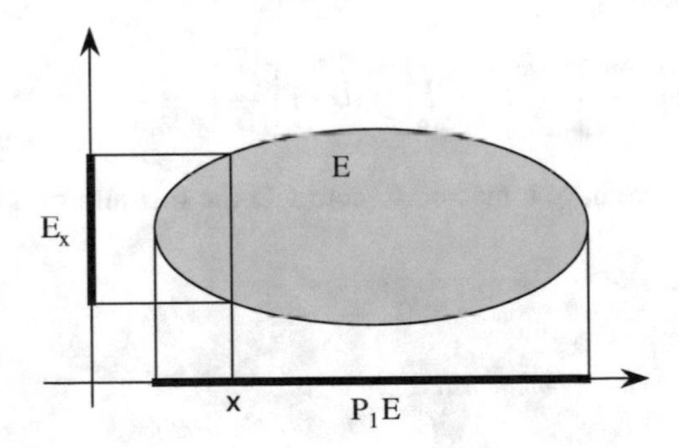

We can then reformulate the Reduction Theorem in the following way.

Theorem 2.34
Let $f : E \to \mathbb{R}$ be an integrable function on the bounded set E. Then:
(i) *for almost every $x \in P_1 E$, the function $f_E(x, \cdot)$ is integrable on the set E_x;*
(ii) *the function $x \mapsto \int_{E_x} f(x, y)\,dy$, defined almost everywhere on $P_1 E$, is integrable there;*
(iii) *we have:*

$$\int_E f = \int_{P_1 E} \left(\int_{E_x} f(x, y)\,dy \right) dx\,.$$

(Continued)

Theorem 2.34 (continued)

Analogously, the function $y \mapsto \int_{E_y} f(x, y)\,dx$*, defined almost everywhere on* P_2E*, is integrable there, and*

$$\int_E f = \int_{P_2E}\left(\int_{E_y} f(x, y)\,dx\right)dy\,.$$

Example Consider the function $f(x, y) = |xy|$ on the set

$$E = \{(x, y) \in \mathbb{R}^2 : 0 \le x \le 1, -x^2 \le y \le x^2\}\,.$$

Being f continuous and E compact, the theorem applies; we have $P_1E = [0, 1]$ and, for every $x \in P_1E$, $E_x = [-x^2, x^2]$. Hence:

$$\begin{aligned}\int_E f &= \int_0^1\left(\int_{-x^2}^{x^2} |xy|\,dy\right)dx\\ &= \int_0^1 |x|\left[\frac{y|y|}{2}\right]_{-x^2}^{x^2} dx = \int_0^1 x^5\,dx = \left[\frac{x^6}{6}\right]_0^1 = \frac{1}{6}\,.\end{aligned}$$

As a corollary, we have a method to compute the measure of a bounded measurable set.

Corollary 2.35

If $E \subseteq \mathbb{R}^2$ *is a bounded and measurable set, then:*

(*i*) *for almost every* $x \in P_1E$*, the set* E_x *is measurable;*

(*ii*) *the function* $x \mapsto \mu(E_x)$*, defined almost everywhere on* P_1E*, is integrable there;*

(*iii*) *we have:*

$$\mu(E) = \int_{P_1E} \mu(E_x)\,dx\,.$$

Analogously, the function $y \mapsto \mu(E_y)$*, defined almost everywhere on* P_2E*, is integrable there, and*

$$\mu(E) = \int_{P_2E} \mu(E_y)\,dy\,.$$

Example Let us compute the area of the disk with radius $R > 0$: let $E = \{(x, y) \in \mathbb{R}^2 : x^2 + y^2 \leq R^2\}$. Being E a compact set, it is measurable. We have that $P_1 E = [-R, R]$ and, for every $x \in P_1 E$, $E_x = [-\sqrt{R^2 - x^2}, \sqrt{R^2 - x^2}]$. Hence:

$$\begin{aligned} \mu(E) &= \int_{-R}^{R} 2\sqrt{R^2 - x^2}\, dx = \int_{-\pi/2}^{\pi/2} 2R^2 \cos^2 t\, dt \\ &= R^2 [t + \cos t \sin t]_{-\pi/2}^{\pi/2} = \pi R^2 . \end{aligned}$$

In the case of functions of more than two variables, analogous results to the preceding ones hold true, with the same proofs. One simply needs to separate the variables in two different groups, calling x the first group and y the second one, and the same formulas hold true.

Example We want to compute the volume of the three-dimensional ball with radius $R > 0$: let $E = \{(x, y, z) \in \mathbb{R}^3 : x^2 + y^2 + z^2 \leq R^2\}$. Let us group together the variables (y, z) and consider the projection on the x-axis: $P_1 E = [-R, R]$. The sections E_x then are disks of radius $\sqrt{R^2 - x^2}$, and we have:

$$\mu(E) = \int_{-R}^{R} \pi(R^2 - x^2)\, dx = \pi(2R^3) - \pi \left[\frac{x^3}{3}\right]_{-R}^{R} = \frac{4}{3}\pi R^3 .$$

Another way to compute the same volume is to group the variables (x, y) and consider $P_1 E = \{(x, y) : x^2 + y^2 = R^2\}$. For every $(x, y) \in P_1 E$, it is

$$E_{(x,y)} = \left[-\sqrt{R^2 - x^2 - y^2}, \sqrt{R^2 - x^2 - y^2}\right],$$

so that

$$\begin{aligned} \mu(E) &= \int_{P_1 E} 2\sqrt{R^2 - x^2 - y^2}\, dx\, dy = \int_{-R}^{R} \left(\int_{-\sqrt{R^2-x^2}}^{\sqrt{R^2-x^2}} 2\sqrt{R^2 - x^2 - y^2}\, dy \right) dx \\ &= \int_{-R}^{R} \left(\int_{-\pi/2}^{\pi/2} 2(R^2 - x^2) \cos^2 t\, dt \right) dx = \int_{-R}^{R} \pi(R^2 - x^2)\, dx = \frac{4}{3}\pi R^3 , \end{aligned}$$

by the change of variable $t = \arcsin\left(y/\sqrt{R^2 - x^2}\right)$.

Iterating the above reduction procedure, it is possible to prove, for a function of N variables which is integrable on a rectangle

$$I = [a_1, b_1] \times [a_2, b_2] \times \cdots \times [a_N, b_N] ,$$

formulas like

$$\int_I f = \int_{a_1}^{b_1} \left(\int_{a_2}^{b_2} \left(\dots \int_{a_N}^{b_N} f(x_1, x_2, \dots, x_N) \, dx_N \, \dots \right) dx_2 \right) dx_1 \, .$$

Exercises

1. Compute the following integrals:

$$\int_{[0,1]\times[2,3]} x^2 y \, dx \, dy \, ,$$

$$\int_{[-1,1]\times[1,2]} y \, e^{xy^2} \, dx \, dy \, ,$$

$$\int_{[-1,1]\times[-2,2]} \sin(x^3) \arctan(x^2 y) \, dx \, dy \, .$$

2. Compute, for any $\alpha \geq 1$, the area of the set

$$E = \{(x, y) \in \mathbb{R}^2 : 1 \leq x \leq \alpha, \; x^2 - y^2 \geq 1\} \, .$$

3. Compute the integral

$$\int_E \sin^2(xy) \, dx \, dy \, ,$$

with the same set E defined above.

4. Compute the integral

$$\int_E (x^2 + y^2) \, dx \, dy \, ,$$

where $E = \{(x, y) \in \mathbb{R}^2 : x^2 + y^2 \leq R^2\}$, for some $R > 0$.

5. Compute the volume of the set

$$\left\{ (x, y, z) \in \mathbb{R}^3 : \frac{x^2}{a^2} + \frac{y^2}{b^2} + \frac{z^2}{c^2} \leq 1 \right\} ,$$

where a, b, c are positive constants (geometrically corresponding to the three semi-axes of an ellipsoid).

2.10 Change of Variables in the Integral

In this section we look for an analogue to the formula of integration by substitution, which was proved in ► Chap. 1 for functions of a single variable. The proof of that formula was based on the Fundamental Theorem. Since we do not have such a powerful tool for functions of several variables, actually we will not be able to completely generalize that formula.

The function φ will be not only assumed to be differentiable, but we will need it to be a diffeomorphism between two open sets A and B of $\mathbb{R}^N$. In other words, $\varphi : A \to B$ will be continuously differentiable, invertible, and $\varphi^{-1} : B \to A$ continuously differentiable, as well. It is useful to recall that a diffeomorphism transforms open sets into open sets, closed sets into closed sets and, for every point $q \in A$, the Jacobian matrix $\varphi'(q)$ is invertible: it is $\det \varphi'(q) \neq 0$. Moreover, we will need the following property.

Lemma 2.36
Let $A \subseteq \mathbb{R}^N$ be an open set, and $\varphi : A \to \mathbb{R}^N$ be a C^1-function; if S is a subset of A of the type

$$S = [a_1, b_1] \times \cdots \times [a_{N-1}, b_{N-1}] \times \{c\},$$

then $\varphi(S)$ is negligible.

Proof
For simplicity, let us concentrate on the case of a subset of $\mathbb{R}^2$ of the type

$$S = [0, 1] \times \{0\}.$$

Consider the rectangles (indeed squares)

$$J_{k,n} = \left[\frac{k-1}{n}, \frac{k}{n}\right] \times \left[-\frac{1}{2n}, \frac{1}{2n}\right],$$

with $k = 1, \ldots, n$. For n large enough, they are contained in a rectangle R which is itself contained in A. Being R a compact set, there is a constant $C > 0$ such that $\|\varphi'(q)\| \leq C$, for every $q \in R$. Then, φ is Lipschitz continuous on R with Lipschitz constant C. Since the sets $J_{k,n}$ have as diameter $\frac{1}{n}\sqrt{2}$, the sets $\varphi(J_{k,n})$ are surely contained in some squares $\tilde{J}_{k,n}$ whose sides' lengths are equal to $\frac{C}{n}\sqrt{2}$. We then have that $\varphi(S)$ is covered by the rectangles $\tilde{J}_{k,n}$, and

$$\sum_{k=1}^{n} \mu(\tilde{J}_{k,n}) \leq n \left(\frac{C}{n}\sqrt{2}\right)^2 = \frac{2C^2}{n}.$$

Since this quantity can be made arbitrarily small, the conclusion follows from Corollary 2.22.
□

As a consequence of the above lemma, it is easy to see that the image of the boundary of a rectangle through a diffeomorphism φ is a negligible set. In particular, given two non-overlapping rectangles, their images are non-overlapping sets.

We are now ready to prove a first version of the **Theorem on the Change of Variables** in the integral, which will be generalized in a later section.

Theorem 2.37

Let φ be a diffeomorphism between two open and bounded sets A and $B = \varphi(A)$, and $f : B \to \mathbb{R}$ be a continuous function. Then, for every closed subset D of A, we have:

$$\int_{\varphi(D)} f(\boldsymbol{x})\, d\boldsymbol{x} = \int_D f(\varphi(\boldsymbol{u}))\, |\det \varphi'(\boldsymbol{u})|\, d\boldsymbol{u}\,.$$

Proof

Notice first of all that the integrals appearing in the formula are both meaningful, being the sets D and $\varphi(D)$ compact and the considered functions continuous. We will proceed by induction on the dimension N. Let us first consider the case $N = 1$.

First of all, using the method of integration by substitution, one verifies that the formula is true when D is a compact interval $[a, b]$: it is sufficient to consider the two possible cases in which φ is increasing or decreasing, and recall that every continuous function is primitivable. For instance, if φ is decreasing, we have $\varphi([a, b]) = [\varphi(b), \varphi(a)]$, so that:

$$\begin{aligned}\int_{\varphi([a,b])} f(x)\, dx &= \int_{\varphi(b)}^{\varphi(a)} f(x)\, dx \\ &= \int_b^a f(\varphi(u))\varphi'(u)\, du \\ &= \int_a^b f(\varphi(u))|\varphi'(u)|\, du \\ &= \int_{[a,b]} f(\varphi(u))|\varphi'(u)|\, du\,.\end{aligned}$$

Let now R be a closed subset of A whose interior $\mathring{R}$ contains D. Being both f and $(f \circ \varphi)|\varphi'|$ continuous, they are integrable on the compact sets $\varphi(R)$ and R, respectively. The open sets $\mathring{R}$ and $\mathring{R} \setminus D$ can each be split into a countable union of non-overlapping compact intervals, whose images through φ also are non-overlapping close intervals. By the complete additivity of the integral, the formula holds true for $\mathring{R}$ and $\mathring{R} \setminus D$:

$$\int_{\varphi(\mathring{R})} f(x)\, dx = \int_{\mathring{R}} f(\varphi(u))|\varphi'(u)|\, du\,,$$

$$\int_{\varphi(\mathring{R}\setminus D)} f(x)\, dx = \int_{\mathring{R}\setminus D} f(\varphi(u))|\varphi'(u)|\, du\,.$$

Hence,

$$\begin{aligned}\int_{\varphi(D)} f(x)\,dx &= \int_{\varphi(\mathring{R}\setminus(\mathring{R}\setminus D))} f(x)\,dx \\ &= \int_{\varphi(\mathring{R})} f(x)\,dx - \int_{\varphi(\mathring{R}\setminus D)} f(x)\,dx \\ &= \int_{\mathring{R}} f(\varphi(u))|\varphi'(u)|\,du - \int_{\mathring{R}\setminus D} f(\varphi(u))|\varphi'(u)|\,du \\ &= \int_D f(\varphi(u))|\varphi'(u)|\,du\,,\end{aligned}$$

so that the formula is proved in the case $N = 1$.

Assume now that the formula holds for the dimension N, and let us prove that it also holds for $N + 1$.[2] Once we fix a point $\bar{\boldsymbol{u}} \in A$, at least one of the partial derivatives $\frac{\partial \varphi_i}{\partial u_j}(\bar{\boldsymbol{u}})$ is non-zero. We can assume without loss of generality that it is $\frac{\partial \varphi_{N+1}}{\partial u_{N+1}}(\bar{\boldsymbol{u}}) \neq 0$. Consider the function

$$\alpha(u_1, \ldots, u_{N+1}) = (u_1, \ldots, u_N, \varphi_{N+1}(u_1, \ldots, u_{N+1}))\,.$$

Being $\det \alpha'(\bar{\boldsymbol{u}}) = \frac{\partial \varphi_{N+1}}{\partial u_{N+1}}(\bar{\boldsymbol{u}}) \neq 0$, we have that α is a diffeomorphism between an open neighborhood U of $\bar{\boldsymbol{u}}$ and an open neighborhood V of $\alpha(\bar{\boldsymbol{u}})$. Assume first that D be contained in U, and set $\widetilde{D} = \alpha(D)$.

We define on V the function $\beta = \varphi \circ \alpha^{-1}$, which is of the form

$$\beta(v_1, \ldots, v_{N+1}) = (\beta_1(v_1, \ldots, v_{N+1}), \ldots, \beta_N(v_1, \ldots, v_{N+1}), v_{N+1})\,,$$

where, for $j = 1, \ldots, N$, it is

$$\beta_j(v_1, \ldots, v_{N+1}) = \varphi_j(v_1, \ldots, v_N, [\varphi_{N+1}(v_1, \ldots, v_N, \cdot)]^{-1}(v_{N+1}))\,.$$

Such a function β is a diffeomorphism between the open sets V and $W = \varphi(U)$.

Consider the sections

$$V_t = \{(v_1, \ldots, v_N) : (v_1, \ldots, v_N, t) \in V\}\,,$$

and the projection

$$P_{N+1}V = \{t : V_t \neq \emptyset\}\,.$$

For $t \in P_{N+1}V$, define the function

$$\beta_t(v_1, \ldots, v_N) = (\beta_1(v_1, \ldots, v_N, t), \ldots, \beta_N(v_1, \ldots, v_N, t))\,,$$

[2] At a first reading, it is advisable to consider the transition from $N = 1$ to $N + 1 = 2$.

which happens to be a diffeomorphism, defined on the open set V_t, whose image is the open set

$$W_t = \{(x_1, \dots, x_N) : (x_1, \dots, x_N, t) \in W\}.$$

Moreover, $\det \beta_t'(v_1, \dots, v_N) = \det \beta'(v_1, \dots, v_N, t)$. Consider also the sections

$$\widetilde{D}_t = \{(v_1, \dots, v_N) : (v_1, \dots, v_N, t) \in \widetilde{D}\},$$

and the projection

$$P_{N+1}\widetilde{D} = \{t : \widetilde{D}_t \neq \emptyset\}.$$

Analogously, we consider $\beta(\widetilde{D})_t$ and $P_{N+1}\beta(\widetilde{D})$. By the definition of β, it is

$$\beta(\widetilde{D})_t = \beta_t(\widetilde{D}_t), \qquad P_{N+1}\beta(\widetilde{D}) = P_{N+1}\widetilde{D}.$$

Using the Reduction Theorem and the inductive assumption, we have:

$$\begin{aligned}
\int_{\beta(\widetilde{D})} f &= \int_{P_{N+1}\beta(\widetilde{D})} \left(\int_{\beta_t(\widetilde{D}_t)} f(x_1, \dots, x_N, t)\, dx_1 \dots dx_N \right) dt \\
&= \int_{P_{N+1}\widetilde{D}} \left(\int_{\widetilde{D}_t} f(\beta_t(v_1, \dots, v_N), t)\, |\det \beta_t'(v_1, \dots, v_N)|\, dv_1 \dots dv_N \right) dt \\
&= \int_{P_{N+1}\widetilde{D}} \left(\int_{\widetilde{D}_t} f(\beta(v_1, \dots, v_N, t))\, |\det \beta'(v_1, \dots, v_N, t)|\, dv_1 \dots dv_N \right) dt \\
&= \int_{\widetilde{D}} f(\beta(\boldsymbol{v}))\, |\det \beta'(\boldsymbol{v})|\, d\boldsymbol{v}.
\end{aligned}$$

Consider now the function $\tilde{f} : V \to \mathbb{R}$ defined as

$$\tilde{f}(\boldsymbol{v}) = f(\beta(\boldsymbol{v}))\, |\det \beta'(\boldsymbol{v})|.$$

Define the sections

$$D_{u_1,\dots,u_N} = \{u_{N+1} : (u_1, \dots, u_N, u_{N+1}) \in D\},$$

and the projection

$$P_{1,\dots,N} D = \{(u_1, \dots, u_N) : D_{u_1,\dots,u_N} \neq \emptyset\}.$$

In an analogous way we define $\alpha(D)_{u_1,\dots,u_N}$ and $P_{1,\dots,N}\alpha(D)$. They are all closed sets and, by the definition of α, we have

$$\alpha(D)_{u_1,\dots,u_N} = \varphi_{N+1}(u_1, \dots, u_N, D_{u_1,\dots,u_N}), \qquad P_{1,\dots,N}\alpha(D) = P_{1,\dots,N} D.$$

Moreover, for every $(u_1, \dots, u_N) \in P_{1,\dots,N} D$, the function defined by

$$t \to \varphi_{N+1}(u_1, \dots, u_N, t)$$

is a diffeomorphism of one variable between the open sets $U_{u_1,\dots,u_N}$ and $V_{u_1,\dots,u_N}$, sections of U and V, respectively. Using the Reduction Theorem and the one-dimensional formula of change of variables proved above, we have that

$$\begin{aligned}
\int_{\alpha(D)} \tilde{f} &= \int_{P_{1,\dots,N}\alpha(D)} \left(\int_{\alpha(D)_{u_1,\dots,u_N}} \tilde{f}(v_1, \dots, v_{N+1})\, dv_{N+1} \right) dv_1 \dots dv_N \\
&= \int_{P_{1,\dots,N} D} \left(\int_{\varphi_{N+1}(u_1,\dots,u_N, D_{u_1,\dots,u_N})} \tilde{f}(v_1, \dots, v_{N+1})\, dv_{N+1} \right) dv_1 \dots dv_N \\
&= \int_{P_{1,\dots,N} D} \left(\int_{D_{u_1,\dots,u_N}} \tilde{f}(u_1, \dots, u_N, \varphi_{N+1}(u_1, \dots, u_{N+1})) \cdot \right. \\
&\qquad\qquad \left. \cdot \left| \frac{\partial \varphi_{N+1}}{\partial u_{N+1}}(u_1, \dots, u_{N+1}) \right| du_{N+1} \right) du_1 \dots du_N \\
&= \int_D \tilde{f}(\alpha(\boldsymbol{u}))\, |\det \alpha'(\boldsymbol{u})|\, d\boldsymbol{u}\,.
\end{aligned}$$

Hence, being $\varphi = \beta \circ \alpha$, we have:

$$\begin{aligned}
\int_{\varphi(D)} f(\boldsymbol{x})\, d\boldsymbol{x} &= \int_{\beta(\widetilde{D})} f(\boldsymbol{x})\, d\boldsymbol{x} \\
&= \int_{\widetilde{D}} f(\beta(\boldsymbol{v}))\, |\det \beta'(\boldsymbol{v})|\, d\boldsymbol{v} \\
&= \int_{\alpha(D)} \tilde{f}(\boldsymbol{v})\, d\boldsymbol{v} \\
&= \int_D \tilde{f}(\alpha(\boldsymbol{u}))\, |\det \alpha'(\boldsymbol{u})|\, d\boldsymbol{u} \\
&= \int_D f(\beta(\alpha(\boldsymbol{u})))\, |\det \beta'(\alpha(\boldsymbol{u}))|\, |\det \alpha'(\boldsymbol{u})|\, d\boldsymbol{u} \\
&= \int_D f(\varphi(\boldsymbol{u}))\, |\det \varphi'(\boldsymbol{u})|\, d\boldsymbol{u}\,.
\end{aligned}$$

We have then proved that, for every $\boldsymbol{u} \in A$, there is a $\delta(\boldsymbol{u}) > 0$ such that the thesis holds true when D is contained in $B[\boldsymbol{u}, \delta(\boldsymbol{u})]$. A gauge δ is thus defined on A. By Lemma 2.20, we can now cover A with a countable family (J_k) of non-overlapping rectangles, each contained in a rectangle of the type $B[\boldsymbol{u}, \delta(\boldsymbol{u})]$, so that the formula holds for the closed sets contained in any of these rectangles.

At this point let us consider an arbitrary closed subset D of A. Then, the formula holds for each $D \cap J_k$ and, by the complete additivity of the integral and the fact that the sets

$\varphi(D \cap J_k)$ are non-overlapping (as a consequence of Lemma 2.36), we have:

$$\begin{aligned}\int_{\varphi(D)} f(\boldsymbol{x})\, d\boldsymbol{x} &= \sum_k \int_{\varphi(D \cap J_k)} f(\boldsymbol{x})\, d\boldsymbol{x} \\ &= \sum_k \int_{D \cap J_k} f(\varphi(\boldsymbol{u}))\, |\det \varphi'(\boldsymbol{u})|\, d\boldsymbol{u} \\ &= \int_D f(\varphi(\boldsymbol{u}))\, |\det \varphi'(\boldsymbol{u})|\, d\boldsymbol{u}\,.\end{aligned}$$

The theorem is thus completely proved. □

Remark The formula on the change of variables is often written, setting $\varphi(D) = E$, in the equivalent form

$$\int_E f(\boldsymbol{x})\, d\boldsymbol{x} = \int_{\varphi^{-1}(E)} f(\varphi(\boldsymbol{u}))\, |\det \varphi'(\boldsymbol{u})|\, d\boldsymbol{u}\,.$$

Example Consider the set

$$E = \{(x, y) \in \mathbb{R}^2 : -1 \leq x \leq 1,\ x^2 \leq y \leq x^2 + 1\}\,,$$

and let $f(x, y) = x^2 y$ be a function on it. Defining $\varphi(u, v) = (u, v + u^2)$, we have a diffeomorphism with $\det \varphi'(u, v) = 1$. Being $\varphi^{-1}(E) = [-1, 1] \times [0, 1]$, by the Theorem on the Change of Variables and the use of the Reduction Theorem we have:

$$\int_E x^2 y\, dx\, dy = \int_{-1}^{1} \left(\int_0^1 u^2 (v + u^2)\, dv \right) du = \int_{-1}^{1} \left(\frac{u^2}{2} + u^4 \right) du = \frac{11}{15}\,.$$

2.11 Change of Measure by Diffeomorphisms

In this section we study how the measure is changed by the action of a diffeomorphism.

Theorem 2.38
Let φ be a diffeomorphism between two open and bounded sets A and B. If D is a measurable subset of A, then $\varphi(D)$ is measurable, $|\det \varphi'|$ is integrable on D, and

$$\mu(\varphi(D)) = \int_D |\det \varphi'(\boldsymbol{u})|\, d\boldsymbol{u}\,.$$

Proof
By the preceding theorem, the formula holds true whenever D is closed. Since every open set can be written as the union of a countable family of non-overlapping (closed) rectangles,

by the complete additivity and the fact that A is bounded, the formula holds true even if D is an open set.

Assume now that D is a measurable set whose closure $\overline{D}$ is contained in A. Let R be a closed subset of A whose interior $\mathring{R}$ contains $\overline{D}$. Then, there is a constant $C > 0$ such that $|\det \varphi'(\boldsymbol{u})| \leq C$ for every $\boldsymbol{u} \in R$. By Proposition 2.21, for every $\varepsilon > 0$ there are two finite or countable families (J_k) and (J'_k), each made of non-overlapping rectangles contained in $\mathring{R}$, such that

$$\mathring{R} \setminus \left(\bigcup_k J'_k\right) \subseteq D \subseteq \bigcup_k J_k\,, \quad \mu\left(\left(\bigcup_k J_k\right) \cap \left(\bigcup_k J'_k\right)\right) \leq \varepsilon\,.$$

Since the formula to be proved holds both on the open sets and on the closed sets, it certainly holds on each rectangle J_k and J'_k; then, it holds on $\cup_k J_k$, on $\cup_k J'_k$, and since it holds even on $\mathring{R}$, it has to be true on $\mathring{R} \setminus (\cup_k J'_k)$, as well. We have thus that $\varphi(\cup_k J_k)$ and $\varphi(\mathring{R} \setminus (\cup_k J'_k))$ are measurable,

$$\varphi\left(\mathring{R} \setminus \left(\bigcup_k J'_k\right)\right) \subseteq \varphi(D) \subseteq \varphi\left(\bigcup_k J_k\right),$$

and

$$\begin{aligned}
\mu\left(\varphi\left(\bigcup_k J_k\right)\right) - \mu\left(\varphi\left(\mathring{R} \setminus \left(\bigcup_k J'_k\right)\right)\right) &= \\
&= \int_{\cup_k J_k} |\det \varphi'(\boldsymbol{u})|\, d\boldsymbol{u} - \int_{\mathring{R} \setminus (\cup_k J'_k)} |\det \varphi'(\boldsymbol{u})|\, d\boldsymbol{u} \\
&= \int_{(\cup_k J_k) \cap (\cup_k J'_k)} |\det \varphi'(\boldsymbol{u})|\, d\boldsymbol{u} \\
&\leq C\mu\left(\left(\bigcup_k J_k\right) \cap \left(\bigcup_k J'_k\right)\right) \\
&\leq C\varepsilon\,.
\end{aligned}$$

Taking $\varepsilon = \frac{1}{n}$, we find in this way two sequences $D_n = \cup_k J_{k,n}$ and $D'_n = \mathring{R} \setminus (\cup_k J'_{k,n})$ with the above properties. By Proposition 2.21, we have that $\varphi(D)$ is measurable and $\mu(\varphi(D)) = \lim_n \mu(\varphi(D_n)) = \lim_n \mu(\varphi(D'_n))$. Moreover, since χ_{D_n} converges almost everywhere to χ_D, by the Dominated Convergence Theorem we have:

$$\begin{aligned}
\mu(\varphi(D)) &= \lim_n \mu(\varphi(D_n)) \\
&= \lim_n \int_{D_n} |\det \varphi'(\boldsymbol{u})|\, d\boldsymbol{u} \\
&= \lim_n \int_R |\det \varphi'(\boldsymbol{u})| \chi_{D_n}(\boldsymbol{u})\, d\boldsymbol{u}
\end{aligned}$$

$$= \int_R |\det \varphi'(\boldsymbol{u})| \chi_D(\boldsymbol{u})\, d\boldsymbol{u}$$

$$= \int_D |\det \varphi'(\boldsymbol{u})|\, d\boldsymbol{u}\,.$$

We can now consider the case of an arbitrary measurable set D in A. Being A open we can consider a sequence of non-overlapping rectangles (K_n) contained in A whose union is A. The formula holds for each of the sets $D \cap K_n$, by the above. The complete additivity of the integral and the fact that A is bounded then permit us to conclude. □

Example Consider the set

$$E = \{(x, y) \in \mathbb{R}^2 : x < y < 2x,\ 3x^2 < y < 4x^2\}.$$

One sees that E is measurable, being an open set. Taking

$$\varphi(u, v) = \left(\frac{u}{v}, \frac{u^2}{v}\right),$$

we have a diffeomorphism between the set $D =]1, 2[\times]3, 4[$ and $E = \varphi(D)$. Moreover,

$$\det \varphi'(u, v) = \det \begin{pmatrix} 1/v & -u/v^2 \\ 2u/v & -u^2/v^2 \end{pmatrix} = \frac{u^2}{v^3}\,.$$

Applying the formula on the change of measure and the Reduction Theorem, we have:

$$\mu(E) = \int_1^2 \left(\int_3^4 \frac{u^2}{v^3}\, dv\right) du = \int_1^2 \frac{7}{288} u^2\, du = \frac{49}{864}\,.$$

2.12 The General Theorem on the Change of Variables

We are now interested in generalizing the Theorem on the Change of Variables assuming f not necessarily continuous, but only L-integrable on a measurable set. In order to do this, it will be useful to prove the following important relation between the integral of a function having non-negative values and the measure of its epigraph.

Proposition 2.39

Let E be a bounded and measurable set and $f : E \to \mathbb{R}$ be a bounded function with non-negative values. Let G_f be the set thus defined:

$$G_f = \{(\boldsymbol{x}, t) \in E \times \mathbb{R} : 0 \le t \le f(\boldsymbol{x})\}\,.$$

(Continued)

Proposition 2.39 (continued)
Then, f is integrable on E if and only if G_f is measurable, in which case

$$\mu(G_f) = \int_E f .$$

Proof
Assume first G_f to be measurable. By the Reduction Theorem, since $P_1 G_f = E$, the sections being $(G_f)_{\boldsymbol{x}} = [0, f(\boldsymbol{x})]$, we have that the function $x \mapsto \int_0^{f(\boldsymbol{x})} 1 = f(\boldsymbol{x})$ is integrable on E, and

$$\mu(G_f) = \int_{G_f} 1 = \int_E \left(\int_0^{f(\boldsymbol{x})} 1\, dt \right) d\boldsymbol{x} = \int_E f(\boldsymbol{x})\, d\boldsymbol{x} .$$

Assume now f to be integrable on E. Let $C > 0$ be a constant such that $0 \leq f(\boldsymbol{x}) < C$, for every $\boldsymbol{x} \in E$. Taken a positive integer n, we divide the interval $[0, C]$ in n equal parts and consider, for $j = 1, \dots, n$, the sets

$$E_n^j = \left\{ \boldsymbol{x} \in E : \frac{j-1}{n} C \leq f(\boldsymbol{x}) < \frac{j}{n} C \right\};$$

by Corollary 2.13 they are measurable, non-overlapping and their union is E. We can then define on E the function ψ_n in the following way:

$$\psi_n = \sum_{j=1}^{n} \frac{j}{n} C \chi_{E_n^j} ,$$

and so

$$G_{\psi_n} = \bigcup_{j=1}^{n} \left(E_n^j \times \left[0, \frac{j}{n} C \right] \right) .$$

By Proposition 2.21, it is easy to see that, being the sets E_n^j measurable, such are the sets $E_n^j \times \left[0, \frac{j}{n} C\right]$, too. Consequently, the sets G_{ψ_n} are measurable. Moreover, since

$$G_f = \bigcap_{n \geq 1} G_{\psi_n} ,$$

even G_f is measurable, and the proof is thus completed. □

We are now in the position to prove the second version of the **Theorem on the Change of Variables** in the integral.

Theorem 2.40
Let φ be a diffeomorphism between two bounded and open sets A and $B = \varphi(A)$ of $\mathbb{R}^N$, D a measurable subset of A and $f : \varphi(D) \to \mathbb{R}$ a function. Then, f is L-integrable on $\varphi(D)$ if and only if $(f \circ \varphi) \, |\det \varphi'|$ is L-integrable on D, in which case we have:

$$\int_{\varphi(D)} f(\boldsymbol{x}) \, d\boldsymbol{x} = \int_D f(\varphi(\boldsymbol{u})) \, |\det \varphi'(\boldsymbol{u})| \, d\boldsymbol{u} .$$

Proof
Assume that f be L-integrable on $E = \varphi(D)$. We first consider the case when f is bounded with non-negative values.

Let $C > 0$ be such that $0 \le f(\boldsymbol{x}) < C$, for every $\boldsymbol{x} \in E$. We define the open sets

$$\tilde{A} = A \times]-C, C[\, , \qquad \tilde{B} = B \times]-C, C[\, ,$$

and the function $\tilde{\varphi} : \tilde{A} \to \tilde{B}$ in the following way:

$$\tilde{\varphi}(u_1, \dots, u_n, t) = (\varphi_1(u_1, \dots, u_n), \dots, \varphi_n(u_1, \dots, u_n), t) .$$

This function is a diffeomorphism and $\det \tilde{\varphi}'(\boldsymbol{u}, t) = \det \varphi'(\boldsymbol{u})$, for every $(\boldsymbol{u}, t) \in \tilde{A}$. Let G_f be the epigraph of f :

$$G_f = \{(\boldsymbol{x}, t) \in E \times \mathbb{R} : 0 \le t \le f(\boldsymbol{x})\} .$$

Being f L-integrable and E measurable, by the preceding proposition we have that G_f is a measurable set. Moreover,

$$\tilde{\varphi}^{-1}(G_f) = \{(\boldsymbol{u}, t) \in D \times \mathbb{R} : 0 \le t \le f(\varphi(\boldsymbol{u}))\} .$$

Using the formula on the change of measure and the Reduction Theorem, we have

$$\begin{aligned}
\mu(G_f) &= \int_{\tilde{\varphi}^{-1}(G_f)} |\det \tilde{\varphi}'(\boldsymbol{u}, t)| \, d\boldsymbol{u} \, dt \\
&= \int_{\tilde{\varphi}^{-1}(G_f)} |\det \varphi'(\boldsymbol{u})| \, d\boldsymbol{u} \, dt \\
&= \int_D \left(\int_0^{f(\varphi(\boldsymbol{u}))} |\det \varphi'(\boldsymbol{u})| \, dt \right) d\boldsymbol{u} \\
&= \int_D f(\varphi(\boldsymbol{u})) \, |\det \varphi'(\boldsymbol{u})| \, d\boldsymbol{u} .
\end{aligned}$$

On the other hand, by Proposition 2.39, we have that $\mu(G_f) = \int_{\varphi(D)} f$, and this proves that the formula holds in case f is bounded with non-negative values.

In the case when f is not bounded but still has non-negative values, we consider the functions

$$f_k(\boldsymbol{x}) = \min\{f(\boldsymbol{x}), k\}.$$

For each of them, the formula holds true, and using the Monotone Convergence Theorem one proves that the formula holds for f even in this case.

When f does not have non-negative values, it is sufficient to consider its positive and negative parts, apply for them the formula and then subtract.

In order to obtain the opposite implication, it is sufficient to consider $(f \circ \varphi)\,|\det \varphi'|$ instead of f and φ^{-1} instead of φ, and to apply what has been just proved. □

We recall here the equivalent formula

$$\int_E f(\boldsymbol{x})\,d\boldsymbol{x} = \int_{\varphi^{-1}(E)} f(\varphi(\boldsymbol{u}))\,|\det \varphi'(\boldsymbol{u})|\,d\boldsymbol{u}.$$

2.13 Some Useful Transformations in $\mathbb{R}^2$

There are some transformations which do not change the measure of any measurable set. We consider here some of those which are most frequently used in applications.

Translations We call translation by a given vector $a = (a_1, a_2) \in \mathbb{R}^2$, the transformation defined by

$$\varphi(u, v) = (u + a_1, v + a_2).$$

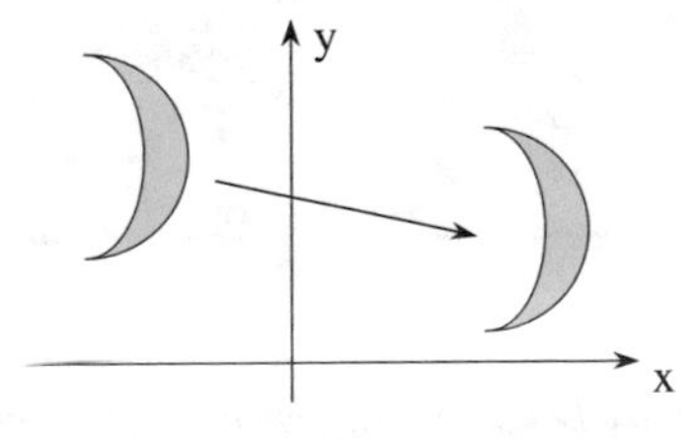

It is readily seen that φ is a diffeomorphism with $\det \varphi' = 1$, so that, given a bounded measurable set D and a L-integrable function f on $\varphi(D)$, we have:

$$\int_{\varphi(D)} f(x, y)\,dx\,dy = \int_D f(u + a_1, v + a_2)\,du\,dv.$$

Reflections A reflection with respect to one of the cartesian axes is defined by

$$\varphi(u,v) = (-u, v)\,, \qquad \text{or} \qquad \varphi(u,v) = (u, -v)\,.$$

Here $\det \varphi' = -1$, so that, taking for example the first case, we have:

$$\int_{\varphi(D)} f(x,y)\,dx\,dy = \int_D f(-u,v)\,du\,dv\,.$$

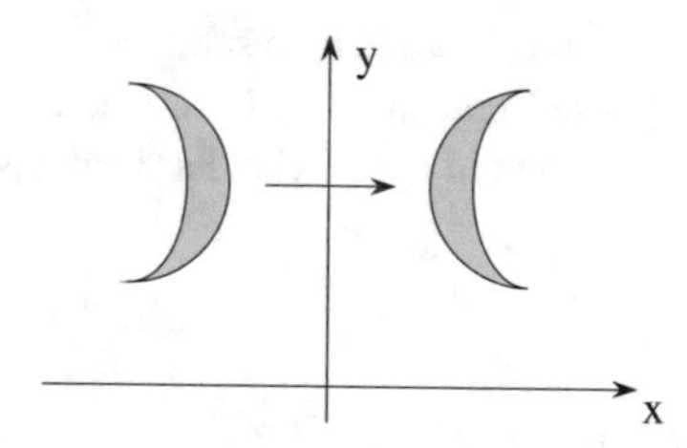

Rotations A rotation around the origin by a fixed angle α is given by

$$\varphi(u,v) = (u\cos\alpha - v\sin\alpha\,,\; u\sin\alpha + v\cos\alpha)\,.$$

It is a diffeomorphism, with

$$\det \varphi'(u,v) = \det \begin{pmatrix} \cos\alpha & -\sin\alpha \\ \sin\alpha & \cos\alpha \end{pmatrix} = (\cos\alpha)^2 + (\sin\alpha)^2 = 1\,.$$

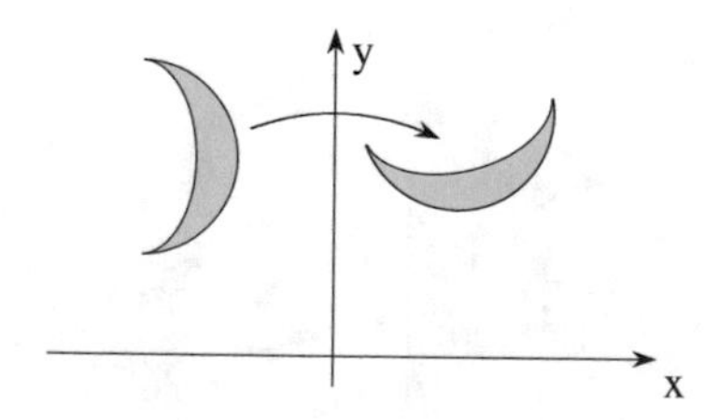

Hence, given a measurable set D and a L-integrable function[3] f on $\varphi(D)$, we have:

$$\int_{\varphi(D)} f(x,y)\,dx\,dy = \int_D f(u\cos\alpha - v\sin\alpha\,,\; u\sin\alpha + v\cos\alpha)\,du\,dv\,.$$

[3]Let us mention here that reference [2] contains an ingenious example of a integrable function in $\mathbb{R}^2$ whose rotation by $\alpha = \pi/4$ is not integrable. This is why we have restricted our attention only to L-integrable functions.

Another useful transformation is the function $\psi : [0, +\infty[\times [0, 2\pi[\to \mathbb{R}^2$ given by

$$\psi(\rho, \theta) = (\rho\cos\theta, \rho\sin\theta),$$

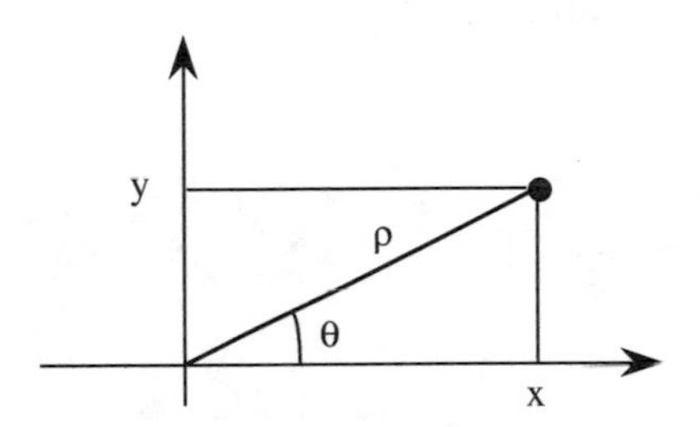

which defines the so-called **polar coordinates** in $\mathbb{R}^2$. Taken a bounded measurable subset of $\mathbb{R}^2$, let B_R be an open ball centered at the origin with radius R which contains it. Consider the open sets

$$A =]0, R[\times]0, 2\pi[\,, \qquad B = B_R \setminus ([0, +\infty[\times \{0\}) .$$

The function $\varphi : A \to B$ defined by $\varphi(\rho, \theta) = \psi(\rho, \theta)$ happens to be a diffeomorphism and it is easily seen that $\det \varphi'(\rho, \theta) = \rho$. We can apply the Theorem on the Change of Variables to the set $\tilde{E} = E \cap B$. Since $\tilde{E}$ and $\varphi^{-1}(\tilde{E})$ differ from E and $\psi^{-1}(E)$, respectively, by negligible sets, we obtain the following **formula on the change of variables in polar coordinates**:

$$\int_E f(x, y)\, dx\, dy = \int_{\psi^{-1}(E)} f(\psi(\rho, \theta))\rho\, d\rho\, d\theta .$$

Example Let $f(x, y) = xy$ be defined on

$$E = \{(x, y) \in \mathbb{R}^2 : x \geq 0, y \geq 0, x^2 + y^2 < 9\} .$$

By the formula on the change of variables in polar coordinates, it is $\psi^{-1}(E) = [0, 3[\times [0, \frac{\pi}{2}]$; by the Reduction Theorem, we can then compute

$$\int_E f = \int_0^{\pi/2} \left(\int_0^3 \rho^3 \cos\theta \sin\theta\, d\rho \right) d\theta = \frac{81}{4} \int_0^{\pi/2} \cos\theta \sin\theta\, d\theta = \frac{81}{8} .$$

2.14 Cylindrical and Spherical Coordinates in $\mathbb{R}^3$

We consider the function $\xi : [0, +\infty[\times [0, 2\pi[\times \mathbb{R} \to \mathbb{R}^3$ defined by

$$\xi(\rho, \theta, z) = (\rho \cos\theta, \rho \sin\theta, z)\,,$$

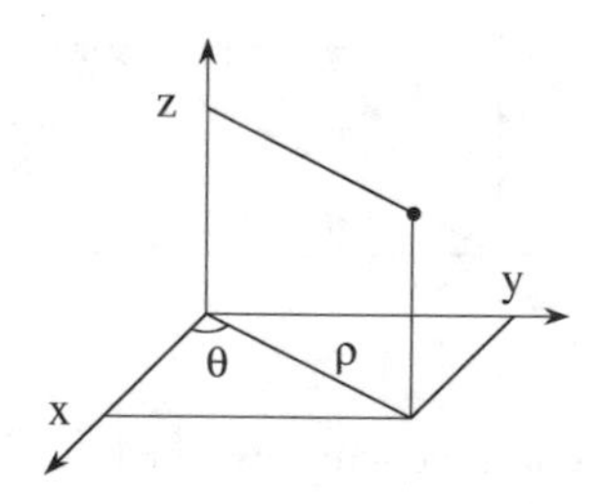

which gives us the so-called **cylindrical coordinates** in $\mathbb{R}^3$. Taken a bounded and measurable set E of $\mathbb{R}^3$, let $C_R \times]-H, H[$ be a cylinder containing it, having as basis the open disk C_R centered at the origin with radius R. Consider the open sets

$$A =]0, R[\times]0, 2\pi[\times]-H, H[\,,$$

$$B = (C_R \setminus ([0, +\infty[\times \{0\})) \times]-H, H[\,.$$

The function $\varphi : A \to B$ defined by $\varphi(\rho, \theta, z) = \xi(\rho, \theta, z)$ happens to be a diffeomorphism and it is easily seen that $\det \varphi'(\rho, \theta, z) = \rho$. We can then apply the Theorem on the Change of Variables to the set $\tilde{E} = E \cap B$. Since $\tilde{E}$ and $\varphi^{-1}(\tilde{E})$ differ from E and $\xi^{-1}(E)$, respectively, by negligible sets, we obtain the following **formula on the change of variables in cylindrical coordinates**:

$$\int_E f(x, y, z)\, dx\, dy\, dz = \int_{\xi^{-1}(E)} f(\xi(\rho, \theta, z))\rho\, d\rho\, d\theta\, dz\,.$$

Example Let us compute the integral $\int_E f$, where $f(x, y, z) = x^2 + y^2$ and

$$E = \{(x, y, z) \in \mathbb{R}^3 : x^2 + y^2 \leq 1, 0 \leq z \leq x + y + \sqrt{2}\}\,.$$

Passing to cylindrical coordinates, we notice that

$$\rho \cos\theta + \rho \sin\theta + \sqrt{2} \geq 0\,,$$

for every $\theta \in [0, 2\pi[$ and every $\rho \in [0, 1]$. By the Theorem on the Change of Variables, using the Reduction Theorem, we compute

$$\begin{aligned}\int_E (x^2+y^2)\,dx\,dy\,dz &= \int_{\xi^{-1}(E)} \rho^3\,d\rho\,d\theta\,dz \\ &= \int_0^1 \left(\int_0^{2\pi}\left(\int_0^{\rho\cos\theta+\rho\sin\theta+\sqrt{2}} \rho^3\,dz\right)d\theta\right)d\rho \\ &= \int_0^1 \left(\int_0^{2\pi} \rho^3(\rho\cos\theta+\rho\sin\theta+\sqrt{2})\,d\theta\right)d\rho \\ &= 2\pi\int_0^1 \rho^3\sqrt{2}\,d\rho \\ &= \frac{\pi\sqrt{2}}{2}\,.\end{aligned}$$

Consider now the function $\sigma : [0, +\infty[\,\times[0, 2\pi[\,\times[0, \pi] \to \mathbb{R}^3$ defined by

$$\sigma(\rho, \theta, \phi) = (\rho\sin\phi\cos\theta, \rho\sin\phi\sin\theta, \rho\cos\phi)\,,$$

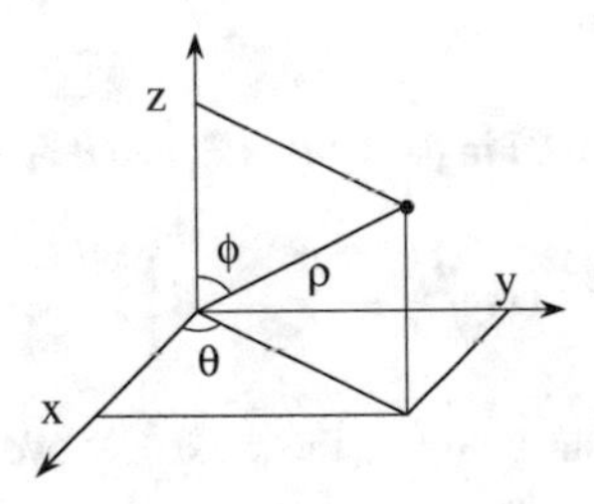

which defines the so-called **spherical coordinates** in $\mathbb{R}^3$. Taken a bounded and measurable subset E of $\mathbb{R}^3$, let B_R be an open three-dimensional ball containing it, centered at the origin with radius R. Consider the open sets

$$A =]0, R[\,\times\,]0, 2\pi[\,\times\,]0, \pi[\,, \qquad B = B_R \setminus ([0, +\infty[\,\times\{0\}\times\mathbb{R})\,.$$

The function $\varphi : A \to B$ defined by $\varphi(\rho, \theta, \phi) = \sigma(\rho, \theta, \phi)$ happens to be a diffeomorphism and it can be easily checked that $\det\varphi'(\rho, \theta, \phi) = \rho^2\sin\phi$. We can then apply the Theorem on the Change of Variables to $\tilde{E} = E \cap B$. Since $\tilde{E}$ and $\varphi^{-1}(\tilde{E})$ differ from E and $\sigma^{-1}(E)$, respectively, by negligible sets, we obtain the following **formula on the change of variables in spherical coordinates**:

$$\int_E f(x, y, z)\,dx\,dy\,dz = \int_{\sigma^{-1}(E)} f(\sigma(\rho, \theta, \phi))\rho^2\sin\phi\,d\rho\,d\theta\,d\phi\,.$$

Example Let us compute the volume of the set

$$E = \left\{(x, y, z) \in \mathbb{R}^3 : x^2 + y^2 + z^2 \leq 1, z \geq \sqrt{x^2 + y^2}\right\}.$$

We have:

$$\begin{aligned}\mu(E) &= \int_E 1 \, dx \, dy \, dz \\ &= \int_{\sigma^{-1}(E)} \rho^2 \sin\phi \, d\rho \, d\theta \, d\phi \\ &= \int_0^1 \left(\int_0^{\pi/4} \left(\int_0^{2\pi} \rho^2 \sin\phi \, d\theta\right) d\phi\right) d\rho \\ &= 2\pi \int_0^1 \left(\int_0^{\pi/4} \rho^2 \sin\phi \, d\phi\right) d\rho \\ &= 2\pi \left(1 - \frac{\sqrt{2}}{2}\right) \int_0^1 \rho^2 \, d\rho \\ &= \left(1 - \frac{\sqrt{2}}{2}\right) \frac{2\pi}{3}.\end{aligned}$$

Exercises

1. Let E be a planar set contained in $[0, +\infty[\times \mathbb{R}$, and define the set

$$E_{\text{rot}} = \left\{(x, y, z) \in \mathbb{R}^3 : (x, \sqrt{y^2 + z^2}) \in E\right\}$$

(i.e., the set obtained rotating E around the x-axis). Prove that

$$\mu(E_{\text{rot}}) = 2\pi \int_E x \, dx \, dy.$$

2. Use the above formula to compute the volume of a sphere: $V = \frac{4}{3}\pi R^3$. Moreover, prove that the volume of the torus with minor radius r and major radius R is equal to $2\pi^2 R r^2$.
3. By some modified cylindrical coordinates, prove that the volume of the cone

$$\left\{(x, y, z) \in \mathbb{R}^3 : \sqrt{\left(\frac{x}{a}\right)^2 + \left(\frac{y}{b}\right)^2} \leq \frac{z}{h} \leq 1\right\},$$

where a, b and h are positive constants, is equal to $\frac{1}{3}(\pi ab)h$.
4. By some modified spherical coordinates, compute again the volume of the ellipsoid

$$\left\{(x, y, z) \in \mathbb{R}^3 : \frac{x^2}{a^2} + \frac{y^2}{b^2} + \frac{z^2}{c^2} \leq 1\right\}.$$

5. Compute the integral

$$\int_{B_1} \sqrt{x^2 + y^2 + z^2}\, dx\, dy\, dz\,,$$

where B_1 is the three-dimensional open ball centered at the origin, with radius 1.

2.15 The Integral on Unbounded Sets

While for L-integrable functions the extension of the theory to unbounded sets does not encounter great difficulties, it seems not to exist a satisfactory general definition of integrability for functions of several variables. This is the reason why, in the following, we will concentrate only on the theory for L-integrable functions. We will use the notation

$$B[0, r] = \left\{(x_1, \ldots, x_N) \in \mathbb{R}^N : \max\{|x_1|, \ldots, |x_N|\} \le r\right\}.$$

Definition 2.41

Given a set $E \subseteq \mathbb{R}^N$, not necessarily bounded, a function $f : E \to \mathbb{R}$ is said to be **L-integrable** (on E) if it is L-integrable on each of the bounded sets $E \cap B[0, r]$, with $r > 0$, and the two following limits exist and are finite:

$$\lim_{r\to+\infty} \int_{E\cap B[0,r]} f\,, \qquad \lim_{r\to+\infty} \int_{E\cap B[0,r]} |f|\,.$$

In this case, the first of these limits is said to be the **integral** of f on E and is denoted by the symbol $\int_E f$.

Equivalently, f is L-integrable on E if the two following limits exist and are finite:

$$\int_E f^+ = \lim_{r\to+\infty} \int_{E\cap B[0,r]} f^+\,, \qquad \int_E f^- = \lim_{r\to+\infty} \int_{E\cap B[0,r]} f^-\,;$$

in that case, we have $\int_E f = \int_E f^+ - \int_E f^-$.

It is not difficult to prove that the set of L-integrable functions is a vector space, and the integral is a linear function on it which preserves the order. Moreover, one easily verifies that a function f is L-integrable on a set E if and only if the function f_E is L-integrable on $\mathbb{R}^N$.

Definition 2.42

A set $E \subseteq \mathbb{R}^N$ is said to be **measurable** if $E \cap B[0, r]$ is measurable, for every $r > 0$. In that case, we set

$$\mu(E) = \lim_{r \to +\infty} \mu(E \cap B[0, r]) .$$

Notice that $\mu(E)$, in some cases, can be $+\infty$. It is finite if and only if the constant function 1 is L-integrable on E, i.e., the characteristic function of E is L-integrable on $\mathbb{R}^N$. The properties of bounded measurable sets extend easily to unbounded sets. In particular, all open sets and all closed sets are measurable.

The **Monotone Convergence Theorem** of B. Levi attains the following general form.

Theorem 2.43

We are given a function f and a sequence of functions f_k, with $k \in \mathbb{N}$, defined almost everywhere on a subset E of $\mathbb{R}^N$, with real values, verifying the following conditions:

1. *the sequence $(f_k)_k$ converges pointwise to f, almost everywhere on E;*
2. *the sequence $(f_k)_k$ is monotone;*
3. *each function f_k is L-integrable on E;*
4. *the real sequence $(\int_E f_k)_k$ has a finite limit.*

Then, f is L-integrable on E, and

$$\int_E f = \lim_{k \to \infty} \int_E f_k .$$

Proof

Assume, for definiteness, that the sequence $(f_k)_k$ is increasing. By considering the sequence $(f_k - f_0)_k$ instead of $(f_k)_k$, we can assume without loss of generality that all the functions have almost everywhere non-negative values. Let $A = \lim_k (\int_E f_k)$; for every $r > 0$, we can apply the Monotone Convergence Theorem on the bounded set $E \cap B[0, r]$, so that f is integrable on $E \cap B[0, r]$ and

$$\int_{E \cap B[0,r]} f = \lim_{k \to \infty} \int_{E \cap B[0,r]} f_k \leq \lim_{k \to \infty} \int_E f_k = A .$$

Let us prove that the limit of $\int_{E \cap B[0,r]} f$ exists, as $r \to +\infty$, and that it is equal to A. Fix $\varepsilon > 0$; there is a $\bar{k} \in \mathbb{N}$ such that, for $k \geq \bar{k}$,

$$A - \frac{\varepsilon}{2} \leq \int_E f_k \leq A ;$$

being moreover

$$\int_E f_{\bar{k}} = \lim_{r\to\infty} \int_{E\cap B[0,r]} f_{\bar{k}} ,$$

there is a $\bar{r} > 0$ such that, for $r \geq \bar{r}$,

$$A - \varepsilon \leq \int_{E\cap B[0,r]} f_{\bar{k}} \leq A .$$

Then, since the sequence $(f_k)_k$ is increasing, we have that, for every $k \geq \bar{k}$ and every $r \geq \bar{r}$, it is

$$A - \varepsilon \leq \int_{E\cap B[0,r]} f_k \leq A .$$

Passing to the limit as $k \to +\infty$, we obtain, for every $r \geq \bar{r}$,

$$A - \varepsilon \leq \int_{E\cap B[0,r]} f \leq A .$$

The proof is thus completed. □

As an immediate consequence there is the analogous statement for the series of functions.

Corollary 2.44
We are given a function f and a sequence of functions f_k, with $k \in \mathbb{N}$, defined almost everywhere on a subset E of $\mathbb{R}^N$, with real values, verifying the following conditions:

1. *the series $\sum_k f_k$ converges pointwise to f, almost everywhere on E;*
2. *for every $k \in \mathbb{N}$ and almost every $\mathbf{x} \in E$, it is $f_k(\mathbf{x}) \geq 0$;*
3. *each function f_k is L-integrable on E;*
4. *the series $\sum_k (\int_E f_k)$ converges.*

Then, f is L-integrable on E and

$$\int_E f = \sum_{k=0}^{\infty} \int_E f_k .$$

From the Monotone Convergence Theorem we deduce, in complete analogy to that seen for bounded sets, the **Dominated Convergence Theorem** of H. Lebesgue.

2

Theorem 2.45
We are given a function f and a sequence of functions f_k, with $k \in \mathbb{N}$, defined almost everywhere on a subset E of $\mathbb{R}^N$, with real values, verifying the following conditions:
1. *the sequence $(f_k)_k$ converges pointwise to f, almost everywhere on E;*
2. *each function f_k is L-integrable on E;*
3. *there are two functions g, h, defined almost everywhere and L-integrable on E, such that*

$$g(\boldsymbol{x}) \leq f_k(\boldsymbol{x}) \leq h(\boldsymbol{x}),$$

for every $k \in \mathbb{N}$ and almost every $\boldsymbol{x} \in E$.

Then, the sequence $(\int_E f_k)_k$ has a finite limit, f is L-integrable on E, and

$$\int_E f = \lim_{k\to\infty} \int_E f_k .$$

As a direct consequence we have the property of **complete additivity of the integral for L-integrable functions**:

Theorem 2.46
Let (E_k) be a finite or countable family of pairwise non-overlapping measurable subsets of $\mathbb{R}^N$, whose union is a set E. Then, f is L-integrable on E if and only if the two following conditions hold:
(a) *f is L-integrable on each set E_k;*
(b) $\sum_k \int_{E_k} |f(\boldsymbol{x})|\,d\boldsymbol{x} < +\infty$.

In that case, we have

$$\int_E f = \sum_k \int_{E_k} f .$$

As another consequence, we have the **Leibniz rule** for not necessarily bounded subsets Y of $\mathbb{R}^N$, which is stated as follows.

Theorem 2.47
Let $f : X \times Y \to \mathbb{R}$ be a function, where X is an interval of $\mathbb{R}$ containing x_0, and Y is a subset of $\mathbb{R}^N$, such that:

(Continued)

Theorem 2.47 (continued)

(*i*) *for every* $x \in X$, *the function* $f(x,\cdot)$ *is L-integrable on* Y, *so that we can define the function*

$$F(x) = \int_Y f(x, \mathbf{y})\, d\mathbf{y}\,;$$

(*ii*) *for every* $x \in X$ *and almost every* $\mathbf{y} \in Y$, *the partial derivative* $\frac{\partial f}{\partial x}(x, \mathbf{y})$ *exists;*

(*iii*) *there are two L-integrable functions* $g, h : Y \to \mathbb{R}$ *such that*

$$g(\mathbf{y}) \le \frac{\partial f}{\partial x}(x, \mathbf{y}) \le h(\mathbf{y})\,,$$

for every $x \in X$ *and almost every* $\mathbf{y} \in Y$.

Then, the function $\frac{\partial f}{\partial x}(x,\cdot)$, *defined almost everywhere on* Y, *is L-integrable there, the derivative of* F *at* x_0 *exists, and we have:*

$$F'(x_0) = \int_Y \left(\frac{\partial f}{\partial x}(x_0, \mathbf{y}) \right) d\mathbf{y}\,.$$

Also the **Reduction Theorem** of G. Fubini extends to functions defined on a not necessarily bounded subset E of $\mathbb{R}^N$. Let $N = N_1 + N_2$ and write $\mathbb{R}^N = \mathbb{R}^{N_1} \times \mathbb{R}^{N_2}$. For every $(\mathbf{x}, \mathbf{y}) \in \mathbb{R}^{N_1} \times \mathbb{R}^{N_2}$, consider the **sections** of E :

$$E_{\mathbf{x}} = \{\mathbf{y} \in \mathbb{R}^{N_2} : (\mathbf{x}, \mathbf{y}) \in E\}\,, \qquad E_{\mathbf{y}} = \{\mathbf{x} \in \mathbb{R}^{N_1} : (\mathbf{x}, \mathbf{y}) \in E\}\,,$$

and the **projections** of E :

$$P_1E = \{\mathbf{x} \in \mathbb{R}^{N_1} : E_{\mathbf{x}} \ne \emptyset\}\,, \qquad P_2E = \{\mathbf{y} \in \mathbb{R}^{N_2} : E_{\mathbf{y}} \ne \emptyset\}\,.$$

We can then reformulate the theorem in the following form.

Theorem 2.48

Let $f : E \to \mathbb{R}$ *be a L-integrable function. Then:*

(*i*) *for almost every* $\mathbf{x} \in P_1E$, *the function* $f(\mathbf{x},\cdot)$ *is L-integrable on the set* $E_{\mathbf{x}}$;

(*ii*) *the function* $\mathbf{x} \mapsto \int_{E_{\mathbf{x}}} f(\mathbf{x}, \mathbf{y})\, d\mathbf{y}$, *defined almost everywhere on* P_1E, *is L-integrable there;*

(*iii*) *we have:*

$$\int_E f = \int_{P_1E} \left(\int_{E_{\mathbf{x}}} f(\mathbf{x}, \mathbf{y})\, d\mathbf{y} \right) d\mathbf{x}\,.$$

(Continued)

2

Theorem 2.48 (continued)
Analogously, the function $\mathbf{y} \mapsto \int_{E_y} f(\mathbf{x}, \mathbf{y})\, d\mathbf{x}$, *defined almost everywhere on* $P_2 E$, *is L-integrable there, and we have:*

$$\int_E f = \int_{P_2 E} \left(\int_{E_y} f(\mathbf{x}, \mathbf{y})\, d\mathbf{x} \right) d\mathbf{y}\,.$$

Proof
Consider for simplicity the case $N_1 = N_2 = 1$, the general case being perfectly analogous. Assume first that f has non-negative values. By the Reduction Theorem for bounded sets, once fixed $r > 0$, we have that, for almost every $x \in P_1 E \cap [-r, r]$, the function $f(x, \cdot)$ is L-integrable on $E_x \cap [-r, r]$; the function $g_r(x) = \int_{E_x \cap [-r,r]} f(x, y)\, dy$, defined almost everywhere on $P_1 E \cap [-r, r]$, is L-integrable there, and

$$\int_{E \cap B[0,r]} f = \int_{P_1 E \cap [-r,r]} g_r(x)\, dx\,.$$

In particular,

$$\int_{P_1 E \cap [-r,r]} g_r(x)\, dx \le \int_E f\,,$$

so that, if $0 < s \le r$, one has that g_r is L-integrable on $P_1 E \cap [-s, s]$, and

$$\int_{P_1 E \cap [-s,s]} g_r(x)\, dx \le \int_E f\,.$$

Keeping s fixed, we let r tend to $+\infty$. Since f has non-negative values, $g_r(x)$ will be increasing with respect to r. Consequently, for almost every $x \in P_1 E \cap [-s, s]$, the limit $\lim_{r\to+\infty} g_r(x)$ exists (possibly infinite), and we set

$$g(x) = \lim_{r\to+\infty} g_r(x) = \lim_{r\to+\infty} \int_{E_x \cap [-r,r]} f(x, y)\, dy\,.$$

Let $T = \{x \in P_1 E \cap [-s, s] : g(x) = +\infty\}$; let us prove that T is negligible. We define the sets

$$E_n^r = \{x \in P_1 E \cap [-s, s] : g_r(x) > n\}.$$

By Theorem 2.12 these are measurable sets and the Chebyshev inequality yields

$$\mu(E_n^r) \le \frac{1}{n} \int_{P_1 E \cap [-s,s]} g_r(x)\, dx \le \frac{1}{n} \int_E f\,.$$

Hence, since the sets E_n^r increase with r, also the sets $F_n = \cup_r E_n^r$ are measurable, and we have that $\mu(F_n) \leq \frac{1}{n} \int_E f$. Being $T \subseteq \cap_n F_n$, we deduce that T is measurable, with $\mu(T) = 0$.

Hence, for almost every $x \in P_1 E \cap [-s, s]$, the function $f(x, \cdot)$ is L-integrable on the set E_x and, by definition,

$$\int_{E_x} f(x, y)\, dy = g(x)\,.$$

Moreover, if we take r in the set of natural numbers and apply the Monotone Convergence Theorem to the functions g_r, it follows that g is L-integrable on $P_1 E \cap [-s, s]$, and

$$\int_{P_1 E \cap [-s,s]} g = \lim_{r \to \infty} \int_{P_1 E \cap [-s,s]} g_r\,,$$

so that

$$\int_{P_1 E \cap [-s,s]} \left(\int_{E_x} f(x, y)\, dy \right) dx \leq \int_E f\,.$$

Letting now s tend to $+\infty$, we see that the limit

$$\lim_{s \to +\infty} \int_{P_1 E \cap [-s,s]} \left(\int_{E_x} f(x, y)\, dy \right) dx$$

exists and in finite; therefore, the function $x \mapsto \int_{E_x} f(x, y)\, dy$, defined almost everywhere on $P_1 E$, is L-integrable there, and its integral is the preceding limit. Moreover, from the above proved inequality, passing to the limit, we have that

$$\int_{P_1 E} \left(\int_{E_x} f(x, y)\, dy \right) dx \leq \int_E f\,.$$

On the other hand,

$$\begin{aligned} \int_{E \cap B[0,r]} f &= \int_{P_1 E \cap [-r,r]} \left(\int_{E_x \cap [-r,r]} f(x, y)\, dy \right) dx \\ &\leq \int_{P_1 E \cap [-r,r]} \left(\int_{E_x} f(x, y)\, dy \right) dx \\ &\leq \int_{P_1 E} \left(\int_{E_x} f(x, y)\, dy \right) dx\,, \end{aligned}$$

so that, passing to the limit as $r \to +\infty$,

$$\int_E f \leq \int_{P_1 E} \left(\int_{E_x} f(x, y)\, dy \right) dx\,.$$

In conclusion, equality must hold, and the proof is thus completed in the case when f has non-negative values. In the general case, just consider f^+ and f^-, and subtract the corresponding formulas. □

The analogous corollary for the computation of the measure holds.

Corollary 2.49

Let E be a measurable set, with a finite measure. Then,

(*i*) *for almost every $\boldsymbol{x} \in P_1 E$, the set $E_{\boldsymbol{x}}$ is measurable and has a finite measure;*

(*ii*) *the function $\boldsymbol{x} \mapsto \mu(E_{\boldsymbol{x}})$, defined almost everywhere on $P_1 E$, is L-integrable there;*

(*iii*) *we have:*

$$\mu(E) = \int_{P_1 E} \mu(E_{\boldsymbol{x}})\, d\boldsymbol{x}\,.$$

With a symmetric statement, if E has a finite measure, we also have

$$\mu(E) = \int_{P_2 E} \mu(E_{\boldsymbol{y}})\, d\boldsymbol{y}\,.$$

The **Theorem on the Change of Variables** also extends to unbounded sets, with the same statement.

Theorem 2.50

Let φ be a diffeomorphism between two open sets A and $B = \varphi(A)$ of $\mathbb{R}^N$, D be a measurable subset of A, and $f : \varphi(D) \to \mathbb{R}$ be a function. Then, f is L-integrable on $\varphi(D)$ if and only if $(f \circ \varphi)\,|\det \varphi'|$ is L-integrable on D, in which case we have:

$$\int_{\varphi(D)} f(\boldsymbol{x})\, d\boldsymbol{x} = \int_D f(\varphi(\boldsymbol{u}))\,|\det \varphi'(\boldsymbol{u})|\, d\boldsymbol{u}\,.$$

Proof

Assume first that f be L-integrable on $E = \varphi(D)$ with non-negative values. Then, for every $r > 0$,

$$\begin{aligned}\int_{D\cap B[0,r]} f(\varphi(\boldsymbol{u}))\,|\det \varphi'(\boldsymbol{u})|\, d\boldsymbol{u} &= \int_{\varphi(D\cap B[0,r])} f(\boldsymbol{x})\, d\boldsymbol{x} \\ &\le \int_{\varphi(D)} f(\boldsymbol{x})\, d\boldsymbol{x}\,,\end{aligned}$$

so that the limit

$$\lim_{r\to+\infty}\int_{D\cap B[0,r]} f(\varphi(\boldsymbol{u}))\,|\det\varphi'(\boldsymbol{u})|\,d\boldsymbol{u}$$

exists and is finite. Then, $(f\circ\varphi)\,|\det\varphi'|$ is L-integrable on D and we have

$$\int_D f(\varphi(\boldsymbol{u}))\,|\det\varphi'(\boldsymbol{u})|\,d\boldsymbol{u} \le \int_{\varphi(D)} f(\boldsymbol{x})\,d\boldsymbol{x}\,.$$

On the other hand, for every $r>0$,

$$\int_{E\cap B[0,r]} f = \int_{\varphi^{-1}(E\cap B[0,r])} (f\circ\varphi)\,|\det\varphi'| \le \int_{\varphi^{-1}(E)} (f\circ\varphi)\,|\det\varphi'|\,,$$

so that, passing to the limit,

$$\begin{aligned}\int_E f(\boldsymbol{x})\,d\boldsymbol{x} &= \lim_{r\to+\infty}\int_{E\cap B[0,r]} f(\boldsymbol{x})\,d\boldsymbol{x}\\ &\le \int_{\varphi^{-1}(E)} f(\varphi(\boldsymbol{u}))\,|\det\varphi'(\boldsymbol{u})|\,d\boldsymbol{u}\,.\end{aligned}$$

The formula is thus proved when f has non-negative values. In general, just proceed as usual, considering f^+ and f^-.

To obtain the opposite implication, it is sufficient to consider $(f\circ\varphi)\,|\det\varphi'|$ instead of f and φ^{-1} instead of φ, and to apply the above. □

Concerning the change of variables in polar coordinates in $\mathbb{R}^2$ or in cylindrical or spherical coordinates in $\mathbb{R}^3$, the same type of considerations we have made for bounded sets extend to the general case, as well.

Example Let $E=\{(x,y)\in\mathbb{R}^2 : x^2+y^2\ge 1\}$ and $f(x,y)=(x^2+y^2)^{-\alpha}$, with $\alpha>0$. We have

$$\begin{aligned}\int_E \frac{1}{(x^2+y^2)^\alpha}\,dx\,dy &= \int_0^{2\pi}\left(\int_1^{+\infty}\frac{1}{\rho^{2\alpha}}\rho\,d\rho\right)d\theta\\ &= 2\pi\int_1^{+\infty}\rho^{1-2\alpha}\,d\rho\,.\end{aligned}$$

It is thus seen that f is integrable on E if and only if $\alpha>1$, in which case the integral is $\frac{\pi}{\alpha-1}$.

Example Let us compute the three-dimensional measure of the set

$$E=\left\{(x,y,z)\in\mathbb{R}^3 : x\ge 1,\ \sqrt{y^2+z^2}\le\frac{1}{x}\right\}.$$

Using Fubini Theorem, grouping together the variables (y, z) we have

$$\mu(E) = \int_1^{+\infty} \pi \frac{1}{x^2}\, dx = \pi\,.$$

Example Consider the function $f(x, y) = e^{-(x^2+y^2)}$, and let us make a change of variables in polar coordinates:

$$\int_{\mathbb{R}^2} e^{-(x^2+y^2)}\, dx\, dy = \int_0^{2\pi} \left(\int_0^{+\infty} e^{-\rho^2} \rho\, d\rho \right) d\theta = 2\pi \left[-\frac{1}{2} e^{-\rho^2} \right]_0^{+\infty} = \pi\,.$$

Notice that, using the Reduction Theorem, we have:

$$\begin{aligned}\int_{\mathbb{R}^2} e^{-(x^2+y^2)}\, dx\, dy &= \int_{-\infty}^{+\infty} \left(\int_{-\infty}^{+\infty} e^{-x^2} e^{-y^2}\, dx \right) dy \\ &= \left(\int_{-\infty}^{+\infty} e^{-x^2}\, dx \right) \left(\int_{-\infty}^{+\infty} e^{-y^2}\, dy \right) \\ &= \left(\int_{-\infty}^{+\infty} e^{-x^2}\, dx \right)^2,\end{aligned}$$

and we thus find again that

$$\int_{-\infty}^{+\infty} e^{-x^2}\, dx = \sqrt{\pi}\,.$$

Exercises

1. Prove that the sets $\mathbb{Q}^2$, $\mathbb{Q} \times \mathbb{R}$ and $\mathbb{R} \times \mathbb{Q}$ are negligible in $\mathbb{R}^2$.
2. Compute the integral

$$\int_{\mathbb{R}^2} \frac{1}{(1 + x^2 + y^2)\sqrt{x^2 + y^2}}\, dx\, dy\,.$$

3. For what values of $\alpha > 0$ is the integral

$$\int_{\mathbb{R}^3} \frac{1}{(1 + x^2 + y^2 + z^2)^\alpha}\, dx\, dy\, dz$$

well defined as a real number? What is its value?

4. Let

$$E_\gamma = \{(x, y) \in \mathbb{R}^2 : x \geq 1,\ 0 \leq y \leq x^\gamma\}\,,$$

for some $\gamma \in \mathbb{R}$. When is the function $f(x, y) = xy$ integrable on E_γ? For those values of γ, compute

$$\int_{E_\gamma} xy\,dx\,dy\,.$$

5. Compute the measure of the four-dimensional ball

$$B_R = \{(x_1, x_2, x_3, x_4) \in \mathbb{R}^4 : x_1^2 + x_2^2 + x_3^2 + x_4^2 \leq R^2\}\,.$$

Differential Forms

Alessandro Fonda

A. Fonda, *The Kurzweil-Henstock Integral for Undergraduates*,
Compact Textbooks in Mathematics,
https://doi.org/10.1007/978-3-319-95321-2_3

In this chapter we develop a theory leading to important extensions to functions of several variables of the formula given by the Fundamental Theorem. Nevertheless, we will not be able to completely generalize that theorem, because we need to assume, for those functions, a somewhat greater regularity.

3.1 The Vector Spaces $\Omega_M(\mathbb{R}^N)$

Consider, for every positive integer M, the sets $\Omega_M(\mathbb{R}^N)$ made by the M-linear antisymmetric functions on $\mathbb{R}^N$, with real values. It is well known that these are vector spaces on $\mathbb{R}$. We also adopt the convention that $\Omega_0(\mathbb{R}^N) = \mathbb{R}$.

If we choose the indices $i_1, \ldots, i_M$ in the set $\{1, \ldots, N\}$, we can define the M-linear antisymmetric function $dx_{i_1,\ldots,i_M}$: it is the function which associates to the vectors

$$\boldsymbol{v}^{(1)} = \begin{pmatrix} v_1^{(1)} \\ \vdots \\ v_N^{(1)} \end{pmatrix}, \ \ldots\ , \ \boldsymbol{v}^{(M)} = \begin{pmatrix} v_1^{(M)} \\ \vdots \\ v_N^{(M)} \end{pmatrix},$$

the real number

$$\det \begin{pmatrix} v_{i_1}^{(1)} & \ldots & v_{i_1}^{(M)} \\ \vdots & \ldots & \vdots \\ v_{i_M}^{(1)} & \ldots & v_{i_M}^{(M)} \end{pmatrix}.$$

Notice that, whenever two indices coincide, we have the zero function. If two indices are exchanged, the function changes sign. Let us recall the following result from elementary algebra.

Proposition 3.1
If $1 \leq M \leq N$, *the space* $\Omega_M(\mathbb{R}^N)$ *has dimension* $\binom{N}{M}$. *A basis is given by* $(dx_{i_1,\ldots,i_M})_{1\leq i_1<\cdots<i_M\leq N}$. *If* $M > N$, *then* $\Omega_M(\mathbb{R}^N) = \{0\}$.

We are mostly interested in the case $N = 3$. Let us give a closer look to the spaces $\Omega_1(\mathbb{R}^3)$, $\Omega_2(\mathbb{R}^3)$ and $\Omega_3(\mathbb{R}^3)$.

Consider $\Omega_1(\mathbb{R}^3)$, the space of linear functions defined on $\mathbb{R}^3$, with values in $\mathbb{R}$. We denote by dx_1, dx_2, dx_3 the following linear functions:

$$dx_1 : \begin{pmatrix} v_1 \\ v_2 \\ v_3 \end{pmatrix} \mapsto v_1\,, \quad dx_2 : \begin{pmatrix} v_1 \\ v_2 \\ v_3 \end{pmatrix} \mapsto v_2\,, \quad dx_3 : \begin{pmatrix} v_1 \\ v_2 \\ v_3 \end{pmatrix} \mapsto v_3\,.$$

the space $\Omega_1(\mathbb{R}^3)$ has dimension 3 and (dx_1, dx_2, dx_3) is one of its bases.

Consider $\Omega_2(\mathbb{R}^3)$, the space of bilinear antisymmetric functions defined on $\mathbb{R}^3 \times \mathbb{R}^3$, with values in $\mathbb{R}$. It has dimension 3, and a basis is given by $(dx_{1,2}, dx_{1,3}, dx_{2,3})$, where

$$dx_{1,2} : \left(\begin{pmatrix} v_1 \\ v_2 \\ v_3 \end{pmatrix}, \begin{pmatrix} v_1' \\ v_2' \\ v_3' \end{pmatrix}\right) \mapsto \det\begin{pmatrix} v_1 & v_1' \\ v_2 & v_2' \end{pmatrix} = v_1 v_2' - v_2 v_1'\,,$$

$$dx_{1,3} : \left(\begin{pmatrix} v_1 \\ v_2 \\ v_3 \end{pmatrix}, \begin{pmatrix} v_1' \\ v_2' \\ v_3' \end{pmatrix}\right) \mapsto \det\begin{pmatrix} v_1 & v_1' \\ v_3 & v_3' \end{pmatrix} = v_1 v_3' - v_3 v_1'\,,$$

$$dx_{2,3} : \left(\begin{pmatrix} v_1 \\ v_2 \\ v_3 \end{pmatrix}, \begin{pmatrix} v_1' \\ v_2' \\ v_3' \end{pmatrix}\right) \mapsto \det\begin{pmatrix} v_2 & v_2' \\ v_3 & v_3' \end{pmatrix} = v_2 v_3' - v_3 v_2'\,.$$

It is useful to recall that

$$dx_{1,1} = dx_{2,2} = dx_{3,3} = 0\,,$$

$$dx_{2,1} = -dx_{1,2}\,, \quad dx_{3,1} = -dx_{1,3}\,, \quad dx_{3,2} = -dx_{2,3}\,.$$

Consider $\Omega_3(\mathbb{R}^3)$, the space of trilinear antisymmetric functions defined on $\mathbb{R}^3 \times \mathbb{R}^3 \times \mathbb{R}^3$, with values in $\mathbb{R}$. We denote by $dx_{1,2,3}$ the following trilinear function:

$$dx_{1,2,3} : \left(\begin{pmatrix} v_1 \\ v_2 \\ v_3 \end{pmatrix}, \begin{pmatrix} v_1' \\ v_2' \\ v_3' \end{pmatrix}, \begin{pmatrix} v_1'' \\ v_2'' \\ v_3'' \end{pmatrix}\right) \mapsto \det\begin{pmatrix} v_1 & v_1' & v_1'' \\ v_2 & v_2' & v_2'' \\ v_3 & v_3' & v_3'' \end{pmatrix}.$$

Every element of the vector space $\Omega_3(\mathbb{R}^3)$ is a scalar multiple of $dx_{1,2,3}$: the space $\Omega_3(\mathbb{R}^3)$ has dimension 1. Recall that

$$dx_{1,2,3} = dx_{2,3,1} = dx_{3,1,2} = -dx_{3,2,1} = -dx_{2,1,3} = -dx_{1,3,2}$$

and, when two indices coincide, we have the zero function.

3.2 Differential Forms in $\mathbb{R}^N$

Definition 3.2

Given an open subset U of $\mathbb{R}^N$, we call **differential form of degree M** (or M-differential form) a function

$$\omega : U \to \Omega_M(\mathbb{R}^N) .$$

If $M \geq 1$, once we consider the basis $(dx_{i_1,\dots,i_M})_{1\leq i_1<\dots<i_M\leq N}$, the components of the M-differential form ω will be denoted by $f_{i_1,\dots,i_M} : U \to \mathbb{R}$. We will then write

$$\omega(\boldsymbol{x}) = \sum_{1\leq i_1<\dots<i_M\leq N} f_{i_1,\dots,i_M}(\boldsymbol{x})\, dx_{i_1,\dots,i_M} .$$

Hence, the M-linear antisymmetric function $\omega(\boldsymbol{x})$ is determined by the $\binom{N}{M}$-dimensional vector

$$F(\boldsymbol{x}) = \big(f_{i_1,\dots,i_M}(\boldsymbol{x})\big)_{1\leq i_1<\dots<i_M\leq N} .$$

A 0-differential form is nothing else than a function defined on U with values in $\mathbb{R}$. We will say that a M-differential form is of class $\mathcal{C}^k$ if all its components are such.

It is possible to define the sum of two M-differential forms: if ω is as above and $\widetilde{\omega}$ is also defined on U and

$$\widetilde{\omega}(\boldsymbol{x}) = \sum_{1\leq i_1<\dots<i_M\leq N} g_{i_1,\dots,i_M}(\boldsymbol{x})\, dx_{i_1,\dots,i_M} ,$$

we define in a natural way $\omega + \widetilde{\omega}$ as follows:

$$(\omega + \widetilde{\omega})(\boldsymbol{x}) = \sum_{1\leq i_1<\dots<i_M\leq N} (f_{i_1,\dots,i_M}(\boldsymbol{x}) + g_{i_1,\dots,i_M}(\boldsymbol{x}))\, dx_{i_1,\dots,i_M} .$$

Moreover, if $c \in \mathbb{R}$, we define $c\,\omega$, the product of the scalar c by the M-differential form ω, in the following way:

$$(c\,\omega)(\boldsymbol{x}) = \sum_{1\leq i_1<\dots<i_M\leq N} c f_{i_1,\dots,i_M}(\boldsymbol{x})\, dx_{i_1,\dots,i_M} .$$

With these definitions, it can be checked that the set of differential forms of degree M becomes a vector space.

Let us give a closer look to the case $N = 3$. Denoting by ω_M a M-differential form, with $M = 1, 2, 3$, we can write

$$\omega_1(\boldsymbol{x}) = f_1(\boldsymbol{x})\, dx_1 + f_2(\boldsymbol{x})\, dx_2 + f_3(\boldsymbol{x})\, dx_3 \,,$$

$$\omega_2(\boldsymbol{x}) = f_{1,2}(\boldsymbol{x})\, dx_{1,2} + f_{1,3}(\boldsymbol{x})\, dx_{1,3} + f_{2,3}(\boldsymbol{x}) dx_{2,3} \,,$$

$$\omega_3(\boldsymbol{x}) = f_{1,2,3}(\boldsymbol{x})\, dx_{1,2,3} \,.$$

Notice that $\omega_1(\boldsymbol{x})$ and $\omega_2(\boldsymbol{x})$ are determined by the three-dimensional vectors

$$F(\boldsymbol{x}) = (f_1(\boldsymbol{x}), f_2(\boldsymbol{x}), f_3(\boldsymbol{x}))\,, \quad \text{and} \quad G(\boldsymbol{x}) = (f_{12}(\boldsymbol{x}), f_{13}(\boldsymbol{x}), f_{23}(\boldsymbol{x}))\,,$$

respectively.

3.3 External Product

Given two differential forms $\omega : U \to \Omega_M(\mathbb{R}^N)$, $\widetilde{\omega} : U \to \Omega_{\tilde{M}}(\mathbb{R}^N)$, of degrees M and $\tilde{M}$, respectively, we want to define the differential form $\omega \wedge \widetilde{\omega}$, of degree $M + \tilde{M}$, which is called **external product** of ω and $\widetilde{\omega}$. If

$$\omega(\boldsymbol{x}) = \sum_{1 \le i_1 < \cdots < i_M \le N} f_{i_1,\ldots,i_M}(\boldsymbol{x})\, dx_{i_1,\ldots,i_M} \,,$$

and

$$\widetilde{\omega}(\boldsymbol{x}) = \sum_{1 \le j_1 < \cdots < j_{\tilde{M}} \le N} g_{j_1,\ldots,j_{\tilde{M}}}(\boldsymbol{x})\, dx_{j_1,\ldots,j_{\tilde{M}}} \,,$$

we set

$$(\omega \wedge \widetilde{\omega})(\boldsymbol{x}) = \sum_{\substack{1 \le i_1 < \cdots < i_M \le N \\ 1 \le j_1 < \cdots < j_{\tilde{M}} \le N}} f_{i_1,\ldots,i_M}(\boldsymbol{x}) g_{j_1,\ldots,j_{\tilde{M}}}(\boldsymbol{x})\, dx_{i_1,\ldots,i_M,j_1,\ldots,j_{\tilde{M}}} \,.$$

Usually the symbol $\wedge$ is omitted when one of the two is a 0-differential form, since the external product is, in this case, similar to the product with a scalar. Notice that, in the above sum, all elements with a repeating index will be zero. Let us see now some properties of the external product.

Proposition 3.3

If $\omega, \tilde{\omega}, \tilde{\tilde{\omega}}$ are three differential forms of degrees $M, \tilde{M}, \tilde{\tilde{M}}$, respectively, then

$$\tilde{\omega} \wedge \omega = (-1)^{M\tilde{M}} \omega \wedge \tilde{\omega}\,,$$

$$(\omega \wedge \tilde{\omega}) \wedge \tilde{\tilde{\omega}} = \omega \wedge (\tilde{\omega} \wedge \tilde{\tilde{\omega}})\,;$$

if $c \in \mathbb{R}$, then

$$(c\,\omega) \wedge \tilde{\omega} = \omega \wedge (c\,\tilde{\omega}) = c(\omega \wedge \tilde{\omega})\,;$$

moreover, when $M = \tilde{M}$,

$$(\omega + \tilde{\omega}) \wedge \tilde{\tilde{\omega}} = (\omega \wedge \tilde{\tilde{\omega}}) + (\tilde{\omega} \wedge \tilde{\tilde{\omega}})\,,$$

$$\tilde{\tilde{\omega}} \wedge (\omega + \tilde{\omega}) = (\tilde{\tilde{\omega}} \wedge \omega) + (\tilde{\tilde{\omega}} \wedge \tilde{\omega})\,.$$

Proof

Assume that ω and $\tilde{\omega}$ are written as above, and let

$$\tilde{\tilde{\omega}}(\boldsymbol{x}) = \sum_{1 \le k_1 < \cdots < k_{\tilde{\tilde{M}}} \le N} h_{k_1,\ldots,k_{\tilde{\tilde{M}}}}(\boldsymbol{x})\, dx_{k_1,\ldots,k_{\tilde{\tilde{M}}}}\,.$$

The first identity is obtained observing that, in order to arrive from the sequence of indices $i_1, \ldots, i_M, j_1, \ldots, j_{\tilde{M}}$ to the one $j_1, \ldots, j_{\tilde{M}}, i_1, \ldots, i_M$, one has first to move j_1 towards the left making M exchanges, then the same has to be done for j_2, if there is one, and so on, till $j_{\tilde{M}}$ is reached. In the total, it is then necessary to operate $M\tilde{M}$ exchanges of indices. Taking into account the fact that the differential form changes sign each time there is an exchange, we have the formula we wanted to prove.

The proof of the second identity (associative property) shows no great difficulties, as well as for the identities where the constant c appears.

Concerning the distributive property, when $M = \tilde{M}$ we have

$$\begin{aligned}
&((\omega + \tilde{\omega}) \wedge \tilde{\tilde{\omega}})(\boldsymbol{x}) = \\
&= \sum_{\substack{1 \le i_1 < \cdots < i_M \le N \\ 1 \le k_1 < \cdots < k_{\tilde{\tilde{M}}} \le N}} (f_{i_1,\ldots,i_M}(\boldsymbol{x}) + g_{i_1,\ldots,i_M}(\boldsymbol{x}))h_{k_1,\ldots,k_{\tilde{\tilde{M}}}}(\boldsymbol{x})\, dx_{i_1,\ldots,i_M,k_1,\ldots,k_{\tilde{\tilde{M}}}} \\
&= \sum_{\substack{1 \le i_1 < \cdots < i_M \le N \\ 1 \le k_1 < \cdots < k_{\tilde{\tilde{M}}} \le N}} (f_{i_1,\ldots,i_M}(\boldsymbol{x})h_{k_1,\ldots,k_{\tilde{\tilde{M}}}}(\boldsymbol{x}) + \\
&\qquad\qquad + g_{i_1,\ldots,i_M}(\boldsymbol{x})h_{k_1,\ldots,k_{\tilde{\tilde{M}}}}(\boldsymbol{x}))\, dx_{i_1,\ldots,i_M,k_1,\ldots,k_{\tilde{\tilde{M}}}} \\
&= ((\omega \wedge \tilde{\tilde{\omega}}) + (\tilde{\omega} \wedge \tilde{\tilde{\omega}}))(\boldsymbol{x})\,.
\end{aligned}$$

The last identity in the statement is proved either in an analogous way, or using the first and the fourth identities. □

If we consider the particular case of the two constant differential forms

$$\omega(\boldsymbol{x}) = dx_1\,, \quad \widetilde{\omega}(\boldsymbol{x}) = dx_2\,, \qquad \text{for every } \boldsymbol{x} \in U,$$

we will have that $(\omega \wedge \widetilde{\omega})(\boldsymbol{x}) = dx_{1,2}$, for every $\boldsymbol{x} \in U$. We can then write

$$dx_1 \wedge dx_2 = dx_{1,2}\,.$$

More generally, in view of the associative property of the external product, we can write

$$dx_{i_1} \wedge \cdots \wedge dx_{i_M} = dx_{i_1,\dots,i_M}\,.$$

In the following, we will use indifferently the one or the other notation.

3.4 External Differential

Given a M-differential form ω if class $\mathcal{C}^1$, we want to define the differential form $d_{ex}\omega$, of degree $M+1$, which is said to be the **external differential** of ω.

If ω is a 0-differential form, $\omega = f : U \to \mathbb{R}$, its external differential $d_{ex}\omega(\boldsymbol{x})$ is just the differential $df(\boldsymbol{x})$, which is a linear function defined on $\mathbb{R}^N$, with values in $\mathbb{R}$. Being, for every $\boldsymbol{v} = (v_1, \dots, v_N)$,

$$df(\boldsymbol{x})\boldsymbol{v} = \frac{\partial f}{\partial x_1}(\boldsymbol{x})\, v_1 + \cdots + \frac{\partial f}{\partial x_N}(\boldsymbol{x})\, v_N\,,$$

we have

$$df(\boldsymbol{x}) = \frac{\partial f}{\partial x_1}(\boldsymbol{x})\, dx_1 + \cdots + \frac{\partial f}{\partial x_N}(\boldsymbol{x})\, dx_N = \sum_{m=1}^{N} \frac{\partial f}{\partial x_m}(\boldsymbol{x})\, dx_m\,.$$

In the general case, if

$$\omega(\boldsymbol{x}) = \sum_{1 \le i_1 < \cdots < i_M \le N} f_{i_1,\dots,i_M}(\boldsymbol{x})\, dx_{i_1} \wedge \cdots \wedge dx_{i_M}\,,$$

we set

$$d_{ex}\omega(\boldsymbol{x}) = \sum_{1 \le i_1 < \cdots < i_M \le N} df_{i_1,\dots,i_M}(\boldsymbol{x}) \wedge dx_{i_1} \wedge \cdots \wedge dx_{i_M}\,,$$

or, equivalently,

$$d_{ex}\omega(\boldsymbol{x}) = \sum_{1\le i_1<\cdots<i_M\le N}\ \sum_{m=1}^{N} \frac{\partial f_{i_1,\ldots,i_M}}{\partial x_m}(\boldsymbol{x})\, dx_m \wedge dx_{i_1} \wedge \cdots \wedge dx_{i_M}\,.$$

In the following, in order to simplify the notations, we will always write $d\omega$ instead of $d_{ex}\omega$. Let us see some properties of the external differential.

Proposition 3.4
If ω and $\widetilde{\omega}$ are two differential forms of class $\mathcal{C}^1$, of degrees M and $\tilde{M}$, respectively, then

$$d(\omega \wedge \widetilde{\omega}) = d\omega \wedge \widetilde{\omega} + (-1)^M \omega \wedge d\widetilde{\omega}\,;$$

if $M = \tilde{M}$ and $c \in \mathbb{R}$, it is

$$d(\omega + \widetilde{\omega}) = d\omega + d\widetilde{\omega}\,,$$

$$d(c\,\omega) = c\,d\omega\,;$$

if ω is of class $\mathcal{C}^2$, then

$$d(d\omega) = 0\,.$$

Proof
Concerning the first identity, if ω and $\widetilde{\omega}$ are as above, we have:

$$\begin{aligned}
d(\omega \wedge \widetilde{\omega})(\boldsymbol{x}) &= \sum_{\substack{1\le i_1<\cdots<i_M\le N\\ 1\le j_1<\cdots<j_{\tilde{M}}\le N}}\ \sum_{m=1}^{N} \frac{\partial}{\partial x_m}(f_{i_1,\ldots,i_M} g_{j_1,\ldots,j_{\tilde{M}}})(\boldsymbol{x})\, dx_{m,i_1,\ldots,i_M,j_1,\ldots,j_{\tilde{M}}}\\
&= \sum_{\substack{1\le i_1<\cdots<i_M\le N\\ 1\le j_1<\cdots<j_{\tilde{M}}\le N}}\ \sum_{m=1}^{N} \bigg(\frac{\partial f_{i_1,\ldots,i_M}}{\partial x_m} g_{j_1,\ldots,j_{\tilde{M}}} +\\
&\qquad\qquad + f_{i_1,\ldots,i_M} \frac{\partial g_{j_1,\ldots,j_{\tilde{M}}}}{\partial x_m}\bigg)(\boldsymbol{x})\, dx_{m,i_1,\ldots,i_M,j_1,\ldots,j_{\tilde{M}}}\\
&= \sum_{\substack{1\le i_1<\cdots<i_M\le N\\ 1\le j_1<\cdots<j_{\tilde{M}}\le N}}\ \sum_{m=1}^{N} \bigg(\frac{\partial f_{i_1,\ldots,i_M}}{\partial x_m} g_{j_1,\ldots,j_{\tilde{M}}}\bigg)(\boldsymbol{x})\, dx_{m,i_1,\ldots,i_M,j_1,\ldots,j_{\tilde{M}}} +\\
&\quad + (-1)^M \sum_{\substack{1\le i_1<\cdots<i_M\le N\\ 1\le j_1<\cdots<j_{\tilde{M}}\le N}}\ \sum_{m=1}^{N} \bigg(f_{i_1,\ldots,i_M} \frac{\partial g_{j_1,\ldots,j_{\tilde{M}}}}{\partial x_m}\bigg)(\boldsymbol{x})\, dx_{i_1,\ldots,i_M,m,j_1,\ldots,j_{\tilde{M}}}\\
&= (d\omega \wedge \widetilde{\omega})(\boldsymbol{x}) + (-1)^M (\omega \wedge d\widetilde{\omega})(\boldsymbol{x})\,.
\end{aligned}$$

The second and third identities follow easily from the linearity of the derivative. Concerning the last identity, we can see that

$$d(d\omega)(x) = \sum_{1\le i_1<\cdots<i_M\le N} \sum_{k=1}^{N}\sum_{m=1}^{N} \frac{\partial}{\partial x_k}\frac{\partial f_{i_1,\dots,i_M}}{\partial x_m}(x)\, dx_{k,m,i_1,\dots,i_M}\,.$$

Since

$$\frac{\partial}{\partial x_k}\frac{\partial f_{i_1,\dots,i_M}}{\partial x_m} = \frac{\partial}{\partial x_m}\frac{\partial f_{i_1,\dots,i_M}}{\partial x_k}\,,$$

taking into account the fact that $dx_k \wedge dx_m = -dx_m \wedge dx_k$, it is seen that all the terms in the sums pairwise eliminate one another, so that $d(d\omega)(x) = 0$. □

3.5 Differential Forms in $\mathbb{R}^3$

When $N = 3$, if ω_1 and $\widetilde{\omega}_1$ are two 1-differential forms, e.g.,

$$\omega_1(x) = f_1(x)\,dx_1 + f_2(x)\,dx_2 + f_3(x)\,dx_3\,,$$
$$\widetilde{\omega}_1(x) = g_1(x)\,dx_1 + g_2(x)\,dx_2 + g_3(x)\,dx_3\,,$$

using the associative and the distributive properties, we have that

$$\omega_1 \wedge \widetilde{\omega}_1 = (f_1g_2 - f_2g_1)\,dx_{1,2} + (f_1g_3 - f_3g_1)\,dx_{1,3} + (f_2g_3 - f_3g_2)\,dx_{2,3}\,.$$

On the other hand, if ω_1 is a 1-differential form and ω_2 is a 2-differential form, e.g.,

$$\omega_1(x) = f_1(x)dx_1 + f_2(x)dx_2 + f_3(x)dx_3\,,$$
$$\omega_2(x) = g_{1,2}(x)\,dx_{1,2} + g_{1,3}(x)\,dx_{1,3} + g_{2,3}(x)\,dx_{2,3}\,,$$

it is

$$\omega_1 \wedge \omega_2 = (f_1g_{2,3} - f_2g_{1,3} + f_3g_{1,2})\,dx_{1,2,3}\,.$$

If we have a 0-differential form $\omega_0 = f : U \to \mathbb{R}$, then

$$d\omega_0(x) = \frac{\partial f}{\partial x_1}(x)\,dx_1 + \frac{\partial f}{\partial x_2}(x)\,dx_2 + \frac{\partial f}{\partial x_3}(x)\,dx_3\,.$$

Taking a 1-differential form

$$\omega_1(x) = f_1(x)\,dx_1 + f_2(x)\,dx_2 + f_3(x)\,dx_3\,,$$

we have

$$d\omega_1(\boldsymbol{x}) = \left(\frac{\partial f_2}{\partial x_1}(\boldsymbol{x}) - \frac{\partial f_1}{\partial x_2}(\boldsymbol{x}) \right) dx_{1,2} + \\ + \left(\frac{\partial f_3}{\partial x_1}(\boldsymbol{x}) - \frac{\partial f_1}{\partial x_3}(\boldsymbol{x}) \right) dx_{1,3} + \\ + \left(\frac{\partial f_3}{\partial x_2}(\boldsymbol{x}) - \frac{\partial f_2}{\partial x_3}(\boldsymbol{x}) \right) dx_{2,3} .$$

If we consider a 2-differential form

$$\omega_2(\boldsymbol{x}) = g_{1,2}(\boldsymbol{x})\, dx_{1,2} + g_{1,3}(\boldsymbol{x})\, dx_{1,3} + g_{2,3}(\boldsymbol{x})\, dx_{2,3} ,$$

then

$$d\omega_2(\boldsymbol{x}) = \left(\frac{\partial g_{2,3}}{\partial x_1}(\boldsymbol{x}) - \frac{\partial g_{1,3}}{\partial x_2}(\boldsymbol{x}) + \frac{\partial g_{1,2}}{\partial x_3}(\boldsymbol{x}) \right) dx_{1,2,3} .$$

At this point, it is time to observe that, in view of future applications, a wiser choice for the basis of the vector space $\Omega_2(\mathbb{R}^3)$ could be the following:

$$\left(dx_{2,3}\, , dx_{3,1}\, , dx_{1,2}\right) .$$

Indeed, in this way, associating,

- to each scalar function $f : U \to \mathbb{R}$,
  either a 0-differential form $\omega_0 = f$,
  or a 3-differential form $\omega_3 = f\, dx_{1,2,3}$;
- to each vector field $F = (F_1, F_2, F_3) : U \to \mathbb{R}^3$,
  either a 1-differential form $\omega_1 = F_1\, dx_1 + F_2\, dx_2 + F_3\, dx_3$,
  or a 2-differential form $\omega_2 = F_1\, dx_{2,3} + F_2\, dx_{3,1} + F_3\, dx_{1,2}$,

we have the following:

$d\omega_0$ corresponds to the **gradient** of f :

$$\text{grad}\, f = \left(\frac{\partial f}{\partial x_1}, \frac{\partial f}{\partial x_2}, \frac{\partial f}{\partial x_3} \right);$$

$d\omega_1$ corresponds to the **curl** of F :

$$\text{curl}\, F = \left(\frac{\partial F_3}{\partial x_2} - \frac{\partial F_2}{\partial x_3}\, , \; \frac{\partial F_1}{\partial x_3} - \frac{\partial F_3}{\partial x_1}\, , \; \frac{\partial F_2}{\partial x_1} - \frac{\partial F_1}{\partial x_2} \right);$$

$d\omega_2$ corresponds to the **divergence** of F :

$$\text{div}\, F = \frac{\partial F_1}{\partial x_1} + \frac{\partial F_2}{\partial x_2} + \frac{\partial F_3}{\partial x_3} .$$

Then, given two vector fields F and $\widetilde{F}$, once we consider the associated 1-differential forms

$$\omega_1 = F_1\,dx_1 + F_2\,dx_2 + F_3\,dx_3\,, \qquad \widetilde{\omega}_1 = \widetilde{F}_1\,dx_1 + \widetilde{F}_2\,dx_2 + \widetilde{F}_3\,dx_3\,,$$

we have that $\omega_1 \wedge \widetilde{\omega}_1$ corresponds to the **vector product** of F and $\widetilde{F}$:

$$F \times \widetilde{F} = (F_2\widetilde{F}_3 - F_3\widetilde{F}_2\,,\ F_3\widetilde{F}_1 - F_1\widetilde{F}_3\,,\ F_1\widetilde{F}_2 - F_2\widetilde{F}_1)\,;$$

if instead of $\widetilde{\omega}_1$ we take the associated 2-differential form

$$\widetilde{\omega}_2 = \widetilde{F}_1\,dx_{2,3} + \widetilde{F}_2\,dx_{3,1} + \widetilde{F}_3\,dx_{1,2}\,,$$

we have that $\omega_1 \wedge \widetilde{\omega}_2$ corresponds to the **scalar product** of F and $\widetilde{F}$:

$$\langle F\,,\widetilde{F}\rangle = F_1\widetilde{F}_1 + F_2\widetilde{F}_2 + F_3\widetilde{F}_3\,.$$

The properties of the external product and those of the external differential lead to formulas involving the gradient, the curl and the divergence. Taking $f : U \to \mathbb{R}$, $\tilde{f} : U \to \mathbb{R}$, $F : U \to \mathbb{R}^3$ and $\widetilde{F} : U \to \mathbb{R}^3$, we have, for example, the following:

$$\begin{aligned}
&\mathrm{curl}(\mathrm{grad}\, f) = 0\,,\\
&\mathrm{div}(\mathrm{curl}\, F) = 0\,,\\
&\mathrm{grad}(f\tilde{f}) = \tilde{f}(\mathrm{grad}\, f) + f(\mathrm{grad}\, \tilde{f})\,,\\
&\mathrm{curl}(fF) = (\mathrm{grad}\, f) \times F + f(\mathrm{curl}\, F)\,,\\
&\mathrm{div}(f\widetilde{F}) = \langle \mathrm{grad}\, f\,, \widetilde{F}\rangle + f(\mathrm{div}\, \widetilde{F})\,,\\
&\mathrm{div}(F \times \widetilde{F}) = \langle \mathrm{curl}\, F\,, \widetilde{F}\rangle - \langle F\,, \mathrm{curl}\, \widetilde{F}\rangle\,.
\end{aligned}$$

The proofs are left to the reader, who might also enjoy the following exercises.

Exercises

1. Let $\omega, \widetilde{\omega} : \mathbb{R}^2 \to \Omega_1(\mathbb{R}^2)$ be defined as

$$\omega(x, y) = y^2\,dx - x^2\,dy\,, \qquad \widetilde{\omega}(x, y) = xy\,dx + (x^2 + y^2)\,dy\,.$$

   Compute $\omega \wedge \widetilde{\omega}$, $d\omega$, $d\widetilde{\omega}$, $d(\omega \wedge \widetilde{\omega})$ and $d\omega \wedge d\widetilde{\omega}$.

2. Let $\omega : \mathbb{R}^3 \to \Omega_2(\mathbb{R}^3)$ be defined as

$$\omega(x, y, z) = x^2yz\,dy \wedge dz + xy^2z\,dz \wedge dx + xyz^2\,dx \wedge dy\,.$$

   Compute $d\omega : \mathbb{R}^3 \to \Omega_3(\mathbb{R}^3)$.

3. Let $f : \mathbb{R}^3 \to \mathbb{R}$ be the function defined as $f(x, y, z) = xy^2z^3$. Compute

$$\Delta f := \mathrm{div}(\mathrm{grad} f)\,.$$

4. Let $F : \mathbb{R}^3 \to \mathbb{R}^3$ be the vector field defined by

$$F(x, y, z) = (x - y + z^2, \ x^2 + yz \ , \ x + y^2 - 2z) .$$

Compute

$$\operatorname{div} F \, , \quad \operatorname{curl} F \, , \quad \operatorname{grad}(\operatorname{div} F) \, , \quad \operatorname{curl}(\operatorname{curl} F) .$$

5. Compute the exterior differential of $\omega : \mathbb{R}^4 \to \Omega_1(\mathbb{R}^4)$, defined as

$$\omega(x_1, x_2, x_3, x_4) = x_2x_3x_4 \, dx_1 + x_1x_3x_4 \, dx_2 + x_1x_2x_4 \, dx_3 + x_1x_2x_3 \, dx_4 .$$

3.6 M-Surfaces

We denote by I a rectangle in $\mathbb{R}^M$, where $1 \le M \le N$.

Definition 3.5

We call **M-surface in** $\mathbb{R}^N$ a function[1] $\sigma : I \to \mathbb{R}^N$ of class $\mathcal{C}^1$. If $M = 1$, σ is also said to be a **curve**; if $M = 2$, we will simply say **surface**. The set $\sigma(I)$ is called the **support** of the M-surface σ. We will say that the M-surface σ is **regular** if, for every $\boldsymbol{u} \in \mathring{I}$, the Jacobian matrix $\sigma'(\boldsymbol{u})$ has rank M.

Consider for example the case $N = 3$. A curve in $\mathbb{R}^3$ is a function $\sigma : [a, b] \to \mathbb{R}^3$, with $\sigma(t) = (\sigma_1(t), \sigma_2(t), \sigma_3(t))$. The curve is regular if, for every $t \in \,]a, b[$, the vector $\sigma'(t) = (\sigma_1'(t), \sigma_2'(t), \sigma_3'(t))$ is not zero. In that case, it is possible to define the following **tangent versor** at the point $\sigma(t)$:

$$\tau_\sigma(t) = \frac{\sigma'(t)}{\|\sigma'(t)\|} .$$

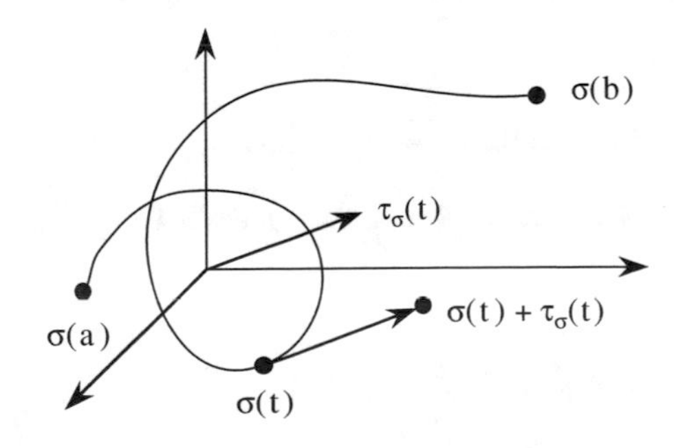

[1] The partial derivatives of σ must be continuous on the whole I, and in the points of the boundary they are interpreted, if necessary, as right or left derivatives. Equivalently, σ could be extended to a $\mathcal{C}^1$-function defined on an open set containing I. In this perspective, the domain of σ could be a more general set than a rectangle, as e.g. the closure of any bounded and open set, so that the differential be well defined at the boundary points, as well. Analogous considerations can be made on the domains of the considered differential forms.

Example The curve $\sigma : [0, 2\pi] \to \mathbb{R}^3$, defined by

$$\sigma(t) = (R\cos(2t), R\sin(2t), 0),$$

has as support the circle

$$\{(x, y, z) : x^2 + y^2 = R^2, z = 0\}$$

(which is covered twice). Being $\sigma'(t) = (-2R\sin(2t), 2R\cos(2t), 0)$, it is a regular curve, and

$$\tau_\sigma(t) = (-\sin(2t), \cos(2t), 0).$$

A surface in $\mathbb{R}^3$ is a function $\sigma : [a_1, b_1] \times [a_2, b_2] \to \mathbb{R}^3$. The surface is regular if, for every $(u, v) \in]a_1, b_1[\times]a_2, b_2[$, the vectors $\frac{\partial\sigma}{\partial u}(u, v)$, $\frac{\partial\sigma}{\partial v}(u, v)$ are linearly independent. In that case, they determine a plane, called the **tangent plane** to the surface at the point $\sigma(u, v)$, and it is possible to define the following **normal versor**:

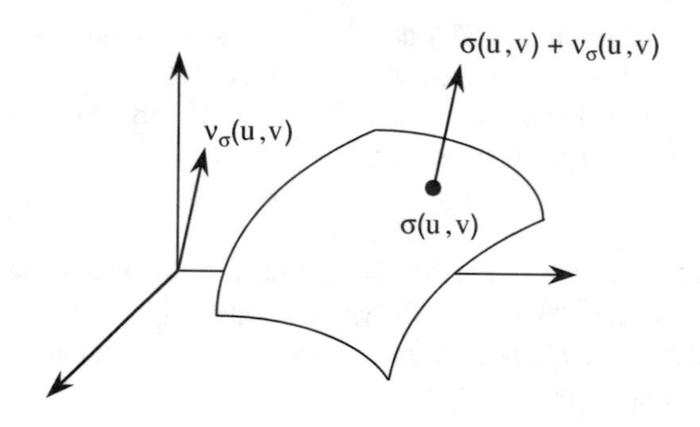

$$\nu_\sigma(u, v) = \frac{\frac{\partial\sigma}{\partial u}(u, v) \times \frac{\partial\sigma}{\partial v}(u, v)}{\|\frac{\partial\sigma}{\partial u}(u, v) \times \frac{\partial\sigma}{\partial v}(u, v)\|}.$$

Examples

1. The surface $\sigma : [0, \pi] \times [0, \pi] \to \mathbb{R}^3$, defined by

   $$\sigma(\phi, \theta) = (R\sin\phi\cos\theta, R\sin\phi\sin\theta, R\cos\phi),$$

   has as support the semi-sphere

   $$\{(x, y, z) : x^2 + y^2 + z^2 = R^2, y \geq 0\}.$$

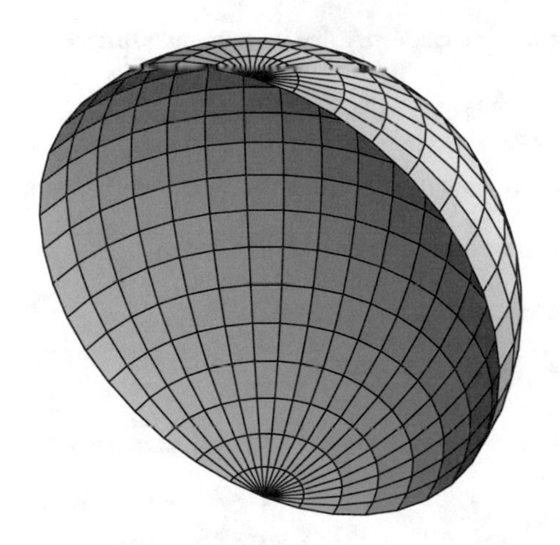

Being

$$\frac{\partial\sigma}{\partial\phi}(\phi,\theta) = (R\cos\phi\cos\theta, R\cos\phi\sin\theta, -R\sin\phi),$$
$$\frac{\partial\sigma}{\partial\theta}(\phi,\theta) = (-R\sin\phi\sin\theta, R\sin\phi\cos\theta, 0),$$

we compute

$$\frac{\partial\sigma}{\partial\phi}(\phi,\theta) \times \frac{\partial\sigma}{\partial\theta}(\phi,\theta) = (R^2\sin^2\phi\cos\theta, R^2\sin^2\phi\sin\theta, R^2\sin\phi\cos\phi).$$

We thus see that it is a regular surface, and

$$\nu_\sigma(\phi,\theta) = (\sin\phi\cos\theta, \sin\phi\sin\theta, \cos\phi).$$

2. The surface $\sigma : [0, 2\pi] \times [0, 2\pi] \to \mathbb{R}^3$, defined by

$$\sigma(u, v) = ((R + r\cos u)\cos v, (R + r\cos u)\sin v, r\sin u),$$

where $0 < r < R$, has as support a torus

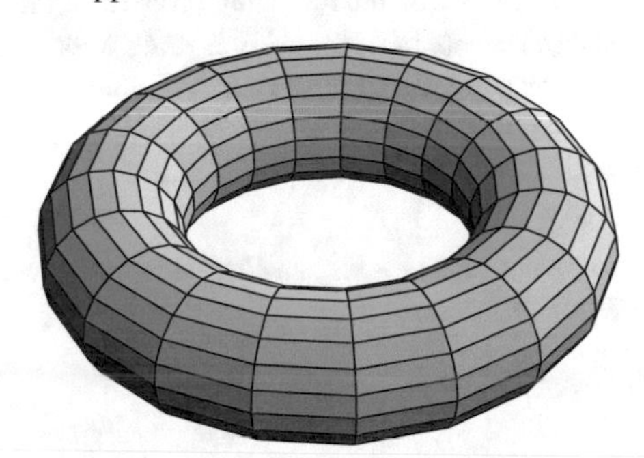

$$\{(x, y, z) : (\sqrt{x^2 + y^2} - R)^2 + z^2 = r^2\}.$$

Even in this case, one can verify that it is a regular surface.

A 3-surface in $\mathbb{R}^3$ is also called a **volume**.

Example The function $\sigma : [0, R] \times [0, \pi] \times [0, 2\pi] \to \mathbb{R}^3$, defined by

$$\sigma(\rho, \phi, \theta) = (\rho \sin\phi \cos\theta, \rho \sin\phi \sin\theta, \rho \cos\phi),$$

has as support the closed ball

$$\{(x, y, z) : x^2 + y^2 + z^2 \le R^2\}.$$

In this case, $\det \sigma'(\rho, \phi, \theta) = \rho^2 \sin\phi$, so that it is a regular volume.

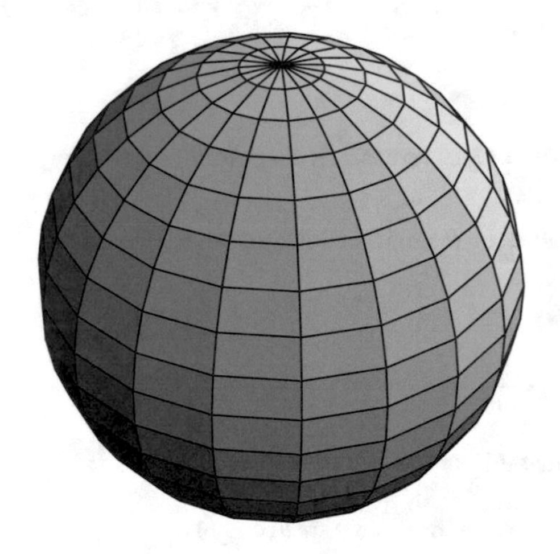

Definition 3.6

Two M-surfaces $\sigma : I \to \mathbb{R}^N$ and $\tilde{\sigma} : J \to \mathbb{R}^N$ are said to be **equivalent** if they have the same support and there are two open sets $A \subseteq I$, $B \subseteq J$, and a diffeomorphism $\varphi : A \to B$ with the following properties: the sets $I \setminus A$ and $J \setminus B$ are negligible and $\sigma(\boldsymbol{u}) = \tilde{\sigma}(\varphi(\boldsymbol{u}))$, for every $\boldsymbol{u} \in A$. We say that σ and $\tilde{\sigma}$ have the **same orientation** if $\det \varphi'(\boldsymbol{u}) > 0$, for every $\boldsymbol{u} \in A$; they have **opposite orientation** if $\det \varphi'(\boldsymbol{u}) < 0$, for every $\boldsymbol{u} \in A$.

Examples Given a curve $\sigma : [a, b] \to \mathbb{R}^N$, an equivalent curve with opposite orientation is, for example, $\tilde{\sigma} : [a, b] \to \mathbb{R}^N$ defined by

$$\tilde{\sigma}(t) = \sigma(a + b - t).$$

If σ is regular, an interesting example of an equivalent curve with the same orientation is obtained by considering the function

$$\varphi(t) = \int_a^t \|\sigma'(\tau)\| \, d\tau \, .$$

Since $\varphi'(t) = \|\sigma'(t)\| > 0$, for every $t \in]a, b[$, setting $\iota_1 = \varphi(b)$, we have that $\varphi : [a, b] \to [0, \iota_1]$ is bijective and the curve $\sigma_1 : [0, \iota_1] \to \mathbb{R}^N$, defined as $\sigma_1(s) = \sigma(\varphi^{-1}(s))$ is equivalent to σ. Notice that, for every $s \in]0, \iota_1[$, it is

$$\begin{aligned} \|\sigma_1'(s)\| &= \|\sigma'(\varphi^{-1}(s))(\varphi^{-1})'(s)\| \\ &= \left\| \sigma'(\varphi^{-1}(s)) \frac{1}{\varphi'(\varphi^{-1}(s))} \right\| \\ &= \left\| \sigma'(\varphi^{-1}(s)) \frac{1}{\|\sigma'(\varphi^{-1}(s))\|} \right\| = 1 \, . \end{aligned}$$

Given a surface $\sigma : [a_1, b_1] \times [a_2, b_2] \to \mathbb{R}^3$, an equivalent surface with opposite orientation is, for example, $\tilde{\sigma} : [a_1, b_1] \times [a_2, b_2] \to \mathbb{R}^3$ defined by

$$\tilde{\sigma}(u, v) = \sigma(u, a_2 + b_2 - v) \, ,$$

or by

$$\tilde{\sigma}(u, v) = \sigma(a_1 + b_1 - u, v) \, .$$

As will be seen in the sequel, two M-surfaces with the same support are not necessarily equivalent. Let us introduce a particular class of M-surfaces for which this inconvenience does not happen.

Definition 3.7

A M-surface $\sigma : I \to \mathbb{R}^N$ is a **M-parametrization** of a set $\mathcal{M}$ if it is regular, injective on $\mathring{I}$, and $\sigma(I) = \mathcal{M}$. We say that a subset of $\mathbb{R}^N$ is **M-parametrizable** if there is a M-parametrization of it.

Examples The circle $\mathcal{M} = \{(x, y) \in \mathbb{R}^2 : x^2 + y^2 = 1\}$ is 1-parametrizable and $\sigma : [0, 2\pi] \to \mathbb{R}^2$, given by $\sigma(t) = (\cos t, \sin t)$, is a 1-parametrization of it.

A 2-parametrization of the sphere $\mathcal{M} = \{(x, y, z) \in \mathbb{R}^3 : x^2 + y^2 + z^2 = 1\}$ is, for example, $\sigma : [0, \pi] \times [0, 2\pi] \to \mathbb{R}^3$, defined by

$$\sigma(\phi, \theta) = (\sin\phi \cos\theta, \sin\phi \sin\theta, \cos\phi) \, .$$

The following theorem is crucial for the treatment of the measure of M-parametrizable M-surfaces.

Theorem 3.8
Two M-parametrizations of the same set are always equivalent.

Proof
Let $\mathcal{M}$ be the subset of $\mathbb{R}^N$ taken in consideration, and let $\sigma : I \to \mathbb{R}^N$ and $\tilde{\sigma} : J \to \mathbb{R}^N$ be two of its M-parametrizations. We define the sets

$$A = \mathring{I} \cap \sigma^{-1}(\mathcal{M} \setminus (\sigma(\partial I) \cup \tilde{\sigma}(\partial J))) , \qquad B = \mathring{J} \cap \tilde{\sigma}^{-1}(\mathcal{M} \setminus (\sigma(\partial I) \cup \tilde{\sigma}(\partial J))) .$$

Then, for every $\boldsymbol{u} \in A$, since $\sigma(\boldsymbol{u}) \in \mathcal{M} \setminus (\sigma(\partial I) \cup \tilde{\sigma}(\partial J))$ and $\tilde{\sigma}(J) = \mathcal{M}$, there exists a $\boldsymbol{v} \in \mathring{J}$ such that $\tilde{\sigma}(\boldsymbol{v}) = \sigma(\boldsymbol{u})$. Clearly, $\tilde{\sigma}(\boldsymbol{v}) \in \mathcal{M} \setminus (\sigma(\partial I) \cup \tilde{\sigma}(\partial J))$, so that $\boldsymbol{v} \in B$. Moreover, since $\tilde{\sigma}$ is injective on $\mathring{J}$, there is a unique $\boldsymbol{v}$ in $\mathring{J}$ with such a property. We can thus define $\varphi : A \to B$ by setting $\varphi(\boldsymbol{u}) = \boldsymbol{v}$. Hence, for $\boldsymbol{u} \in A$ and $\boldsymbol{v} \in B$,

$$\varphi(\boldsymbol{u}) = \boldsymbol{v} \quad \Leftrightarrow \quad \sigma(\boldsymbol{u}) = \tilde{\sigma}(\boldsymbol{v}) .$$

This function $\varphi : A \to B$ is invertible: a symmetrical argument may be used to define its inverse $\varphi^{-1} : B \to A$.

Let us verify that the set A is open. Since $\sigma, \tilde{\sigma}$ are continuous functions and ∂I, ∂J are compact sets, we have that $\sigma(\partial I) \cup \tilde{\sigma}(\partial J)$ is compact, hence closed. Then $\mathcal{M} \setminus (\sigma(\partial I) \cup \tilde{\sigma}(\partial J))$ is relatively open in $\mathcal{M}$, and $\sigma^{-1}(\mathcal{M} \setminus (\sigma(\partial I) \cup \tilde{\sigma}(\partial J))$ is relatively open in I, so that its intersection with $\mathring{I}$ is an open set. In an analogous way it can be seen that B is an open set, as well.

Let us take a $\boldsymbol{v}_0 \in \mathring{J}$, and set $\boldsymbol{x}_0 = \tilde{\sigma}(\boldsymbol{v}_0)$. The Jacobian matrix $\tilde{\sigma}'(\boldsymbol{v}_0)$ has rank M, and we may assume without loss of generality that the first M lines be linearly independent. Being $\mathbb{R}^N \simeq \mathbb{R}^M \times \mathbb{R}^{N-M}$, we will write every point $\boldsymbol{x} \in \mathbb{R}^N$ in the form $\boldsymbol{x} = (\boldsymbol{x}_1, \boldsymbol{x}_2)$, with $\boldsymbol{x}_1 \in \mathbb{R}^M$ and $\boldsymbol{x}_2 \in \mathbb{R}^{N-M}$. However, not to have double indices below, we will write $\boldsymbol{x}_0 = (\boldsymbol{x}_1^0, \boldsymbol{x}_2^0)$.

Let $\Phi : J \times \mathbb{R}^{N-M} \to \mathbb{R}^N$ be defined as

$$\Phi(\boldsymbol{v}, z) = \tilde{\sigma}(\boldsymbol{v}) + (\boldsymbol{0}, z) .$$

Then $\Phi'(\boldsymbol{v}_0, \boldsymbol{0})$ is invertible, so that Φ is a local diffeomorphism: there are an open neighborhood V_0 of $\boldsymbol{v}_0$, an open neighborhood Ω_0 of $\boldsymbol{0}$ in $\mathbb{R}^{N-M}$, and open neighborhood W_0 of $\boldsymbol{x}_0$ such that $\Phi : V_0 \times \Omega_0 \to W_0$ is a diffeomorphism. Moreover, we can assume that $V_0 \subseteq \mathring{J}$. Let $\Psi = \Phi^{-1} : W_0 \to V_0 \times \Omega_0$. We will write $\Psi(\boldsymbol{x}) = (\Psi_1(\boldsymbol{x}), \Psi_2(\boldsymbol{x}))$, with $\Psi_1(\boldsymbol{x}) \in V_0$ and $\Psi_2(\boldsymbol{x}) \in \Omega_0$.

We now prove that φ is of class $\mathcal{C}^1$. Take $\boldsymbol{u}_0 \in A$, and set $\boldsymbol{x}_0 = \sigma(\boldsymbol{u}_0)$ and $\boldsymbol{v}_0 = \varphi(\boldsymbol{u}_0)$. Assume $\boldsymbol{v}_0$ as above, with $\tilde{\sigma}'(\boldsymbol{v}_0)$ having the first M lines linearly independent, so that the local diffeomorphism $\Psi : W_0 \to V_0 \times \Omega_0$ can be defined. Take an open neighborhood U_0 of $\boldsymbol{u}_0$, contained in A, such that $\sigma(U_0) \subseteq W_0$. Then, for $\boldsymbol{u} \in U_0$ and $\boldsymbol{v} \in B$,

$$\varphi(\boldsymbol{u}) = \boldsymbol{v} \quad \Leftrightarrow \quad \sigma(\boldsymbol{u}) = \Phi(\boldsymbol{v}, \boldsymbol{0}) \quad \Leftrightarrow \quad (\boldsymbol{v}, \boldsymbol{0}) = \Psi(\sigma(\boldsymbol{u})) .$$

Hence, φ coincides with $\Psi_1 \circ \sigma$ on the open set U_0, yielding that φ is continuously differentiable.

In a symmetric way it is proved that $\varphi^{-1} : B \to A$ is of class $\mathcal{C}^1$, so that φ happens to be a diffeomorphism.

We now prove that the sets $I \setminus A$ and $J \setminus B$ are negligible. Let us consider, e.g., the second one:

$$J \setminus B = \partial J \cup (\mathring{J} \setminus B) = \partial J \cup \{\boldsymbol{v} \in \mathring{J} : \tilde{\sigma}(\boldsymbol{v}) \in \sigma(\partial I)\} \cup \{\boldsymbol{v} \in \mathring{J} : \tilde{\sigma}(\boldsymbol{v}) \in \tilde{\sigma}(\partial J)\}.$$

We know that ∂J is negligible. Let us prove that $\{\boldsymbol{v} \in \mathring{J} : \tilde{\sigma}(\boldsymbol{v}) \in \sigma(\partial I)\}$ is negligible, as well.

Let $\boldsymbol{v}_0 \in \mathring{J}$ be such that $\tilde{\sigma}(\boldsymbol{v}_0) \in \sigma(\partial I)$. Then, there is a $\boldsymbol{u}_0 \in \partial I$ such that $\sigma(\boldsymbol{u}_0) = \tilde{\sigma}(\boldsymbol{v}_0)$. We argue as above, and define $\Psi : W_0 \to V_0 \times \Omega_0$. Let U_0 be an open neighborhood of $\boldsymbol{u}_0$ such that $\sigma(U_0 \cap I) \subseteq W_0$. Let us see that

$$\mathring{J} \cap \tilde{\sigma}^{-1}(\sigma(U_0 \cap \partial I)) \subseteq (\Psi_1 \circ \sigma)(U_0 \cap \partial I).$$

Indeed, taking $\boldsymbol{v} \in \mathring{J} \cap \tilde{\sigma}^{-1}(\sigma(U_0 \cap \partial I))$, we have that $\tilde{\sigma}(\boldsymbol{v}) \in \sigma(U_0 \cap \partial I)$. Then, being $\Phi(\boldsymbol{v}, \boldsymbol{0}) = \tilde{\sigma}(\boldsymbol{v})$, we have that $\Psi(\tilde{\sigma}(\boldsymbol{v})) = (\boldsymbol{v}, \boldsymbol{0}) \in V_0 \times \Omega_0$, hence $\boldsymbol{v} \in \Psi_1(\sigma(U_0 \cap \partial I))$, and the inclusion is thus proved. Now, since $\Psi_1 \circ \sigma$ is of class $\mathcal{C}^1$, by Lemma 2.36 we have that $(\Psi_1 \circ \sigma)(U_0 \cap \partial I)$ is negligible. Finally, the conclusion that $\{\boldsymbol{v} \in \mathring{J} : \tilde{\sigma}(\boldsymbol{v}) \in \sigma(\partial I)\}$ is negligible follows from the fact that ∂I is compact, so that it can be covered by a finite number of such open sets as U_0.

It remains to be proved that $\{\boldsymbol{v} \in \mathring{J} : \tilde{\sigma}(\boldsymbol{v}) \in \tilde{\sigma}(\partial J)\}$ is negligible. Let $\boldsymbol{v}_0 \in \mathring{J}$ be such that $\tilde{\sigma}(\boldsymbol{v}_0) \in \tilde{\sigma}(\partial J)$. Then, there is a $\tilde{\boldsymbol{v}}_0 \in \partial J$ such that $\tilde{\sigma}(\tilde{\boldsymbol{v}}_0) = \tilde{\sigma}(\boldsymbol{v}_0)$. Let $\widetilde{V}_0$ be an open neighborhood of $\tilde{\boldsymbol{v}}_0$ such that $\tilde{\sigma}(\widetilde{V}_0 \cap J) \subseteq W_0$. As above, one sees that $\mathring{J} \cap \tilde{\sigma}^{-1}(\tilde{\sigma}(\widetilde{V}_0 \cap \partial J)) \subseteq (\Psi_1 \circ \tilde{\sigma})(\widetilde{V}_0 \cap \partial J)$, showing that $\mathring{J} \cap \tilde{\sigma}^{-1}(\tilde{\sigma}(\widetilde{V}_0 \cap \partial J))$ is negligible. The conclusion is obtained as above, covering ∂J by a finite number of such open sets $\widetilde{V}_0$. □

3.7 The Integral of a Differential Form

We want to define the notion of integral of a M-differential form on a M-surface. Let

$$\omega(\boldsymbol{x}) = \sum_{1 \le i_1 < \cdots < i_M \le N} f_{i_1,\ldots,i_M}(\boldsymbol{x})\, dx_{i_1} \wedge \cdots \wedge dx_{i_M},$$

a M-differential form defined on a subset U of $\mathbb{R}^N$ containing the support of a M-surface $\sigma : I \to \mathbb{R}^N$, with $1 \le M \le N$. We consider, when the indices $i_1, \ldots, i_M$ vary in the set $\{1, \ldots, N\}$, the functions $\sigma_{(i_1,\ldots,i_M)} : I \to \mathbb{R}^M$ defined by

$$\sigma_{(i_1,\ldots,i_M)} : \begin{pmatrix} u_1 \\ \vdots \\ u_M \end{pmatrix} \mapsto \begin{pmatrix} \sigma_{i_1}(u_1, \ldots, u_M) \\ \vdots \\ \sigma_{i_M}(u_1, \ldots, u_M) \end{pmatrix}.$$

Definition 3.9

We say that the M-differential form $\omega : U \to \Omega_M(\mathbb{R}^N)$ is **integrable** on the M-surface $\sigma : I \to U$ if, for every choice of the indices $i_1, \dots, i_M$ in the set $\{1, \dots, N\}$, the function $(f_{i_1,\dots,i_M} \circ \sigma) \det \sigma'_{(i_1,\dots,i_M)}$ is integrable on I. In that case, we set

$$\int_\sigma \omega = \sum_{1 \le i_1 < \dots < i_M \le N} \int_I f_{i_1,\dots,i_M}(\sigma(\boldsymbol{u})) \det \sigma'_{(i_1,\dots,i_M)}(\boldsymbol{u})\, d\boldsymbol{u}\,.$$

For example, ω is surely integrable on σ when all its components are continuous functions. Notice that

$$\sigma'_{(i_1,\dots,i_M)}(\boldsymbol{u}) = \frac{\partial(\sigma_{i_1}, \dots, \sigma_{i_M})}{\partial(u_1, \dots, u_M)}(\boldsymbol{u}) = \begin{pmatrix} \frac{\partial \sigma_{i_1}}{\partial u_1}(\boldsymbol{u}) & \dots & \frac{\partial \sigma_{i_1}}{\partial u_M}(\boldsymbol{u}) \\ \vdots & \dots & \vdots \\ \frac{\partial \sigma_{i_M}}{\partial u_1}(\boldsymbol{u}) & \dots & \frac{\partial \sigma_{i_M}}{\partial u_M}(\boldsymbol{u}) \end{pmatrix}.$$

If we define, for every $\boldsymbol{x} \in U$ and every $\boldsymbol{u} \in I$, the $\binom{N}{M}$-dimensional vectors

$$F(\boldsymbol{x}) = \big(f_{i_1,\dots,i_M}(\boldsymbol{x})\big)_{1 \le i_1 < \dots < i_M \le N}\,,$$

$$\Sigma(\boldsymbol{u}) = \big(\det \sigma'_{(i_1,\dots,i_M)}(\boldsymbol{u})\big)_{1 \le i_1 < \dots < i_M \le N}\,,$$

we have that

$$\int_\sigma \omega = \int_I \langle F(\sigma(\boldsymbol{u}))\,, \Sigma(\boldsymbol{u})\rangle\, d\boldsymbol{u}\,,$$

where $\langle \cdot\,, \cdot \rangle$ denotes here the Euclidean scalar product in $\mathbb{R}^{\binom{N}{M}}$.

It is important to analyze how the integral of a differential form ω changes on two equivalent M-surfaces having the same orientations, or opposite orientations.

Theorem 3.10

Let $\sigma : I \to \mathbb{R}^N$ and $\tilde{\sigma} : J \to \mathbb{R}^N$ be two equivalent M-surfaces. If they have the same orientations, then

$$\int_\sigma \omega = \int_{\tilde{\sigma}} \omega\,;$$

if they have opposite orientations, then

$$\int_\sigma \omega = -\int_{\tilde{\sigma}} \omega\,.$$

Proof
We have a M-differential form of the type

$$\omega(\boldsymbol{x}) = \sum_{1\le i_1<\cdots<i_M\le N} f_{i_1,\dots,i_M}(\boldsymbol{x})\, dx_{i_1} \wedge \cdots \wedge dx_{i_M} .$$

Let $\varphi : A \to B$, be as in the definition of equivalent M-surfaces, such that $\sigma = \tilde{\sigma} \circ \varphi$. By the Theorem on the Change of Variables in the integral, it is

$$\begin{aligned}
\int_\sigma \omega &= \sum_{1\le i_1<\cdots<i_M\le N} \int_A f_{i_1,\dots,i_M}(\tilde{\sigma}(\varphi(\boldsymbol{u}))) \det(\tilde{\sigma}\circ\varphi)'_{(i_1,\dots,i_M)}(\boldsymbol{u})\, d\boldsymbol{u} \\
&= \sum_{1\le i_1<\cdots<i_M\le N} \int_A f_{i_1,\dots,i_M}(\tilde{\sigma}(\varphi(\boldsymbol{u}))) \det \tilde{\sigma}'_{(i_1,\dots,i_M)}(\varphi(\boldsymbol{u})) \det \varphi'(\boldsymbol{u})\, d\boldsymbol{u} \\
&= \pm \sum_{1\le i_1<\cdots<i_M\le N} \int_B f_{i_1,\dots,i_M}(\tilde{\sigma}(\boldsymbol{v})) \det \tilde{\sigma}'_{(i_1,\dots,i_M)}(\boldsymbol{v})\, d\boldsymbol{v} \\
&= \pm \int_{\tilde{\sigma}} \omega ,
\end{aligned}$$

with positive sign if $\det \varphi' > 0$, negative if $\det \varphi' < 0$. □

Remark In general, if σ and $\tilde{\sigma}$ are equivalent, we do not necessarily have the equality $|\int_\sigma \omega| = |\int_{\tilde{\sigma}} \omega|$. It is not guaranteed, indeed that they have the same or opposite orientations. For example, if we consider the two surfaces $\sigma, \tilde{\sigma} : [1, 2] \times [0, 2\pi] \to \mathbb{R}^3$, defined by

$$\sigma(u, v) = \left(\left(\frac{3}{2} + \left(u - \frac{3}{2}\right)\cos\frac{v}{2}\right)\cos v, \left(\frac{3}{2} + \left(u - \frac{3}{2}\right)\cos\frac{v}{2}\right)\sin v, \left(u - \frac{3}{2}\right)\sin\frac{v}{2}\right),$$

$$\tilde{\sigma}(u, v) = \sigma\left(u, v + \frac{\pi}{2}\right),$$

it is possible to see that they are both parametrizations of the same set (a Möbius strip), and therefore they are equivalent (the reader is invited to explicitly find a diffeomorphism $\varphi : A \to B$ with the properties required by the definition). On the other hand, if we consider the 2-differential form $\omega(x_1, x_2, x_3) = dx_{12}$, determined by the constant vector field $(0, 0, 1)$, computation yields

$$\int_\sigma \omega = 0 , \quad \int_{\tilde{\sigma}} \omega = -3\sqrt{2} .$$

We now consider the important case when $M = N$.

Theorem 3.11
Let $M = N$; if σ is regular and injective on $\mathring{I}$ with $\det \sigma' > 0$, and ω is of the type

$$\omega(\boldsymbol{x}) = f(\boldsymbol{x})\, dx_1 \wedge \cdots \wedge dx_N\,,$$

then

$$\int_\sigma \omega = \int_{\sigma(I)} f\,.$$

Proof
Using the theorem of local inversion it is seen that σ induces a diffeomorphism between $\mathring{I}$ and $\sigma(\mathring{I})$. Being both the boundary of I and its image through σ negligible (see Lemma 2.36), by the Theorem on the Change of Variables in the integral, we have

$$\begin{aligned}\int_\sigma \omega &= \int_I f(\sigma(\boldsymbol{u}))\det(\sigma'(\boldsymbol{u}))\, d\boldsymbol{u}\\ &= \int_{\mathring{I}} f(\sigma(\boldsymbol{u}))\det(\sigma'(\boldsymbol{u}))\, d\boldsymbol{u}\\ &= \int_{\sigma(\mathring{I})} f = \int_{\sigma(I)} f\,.\end{aligned}$$

This completes the proof. □

If σ is the identity function, then $\sigma(I) = I$, and instead of $\int_\sigma \omega$ one usually writes $\int_I \omega$. Hence, we have that

$$\int_I f(\boldsymbol{x})\, dx_1 \wedge \cdots \wedge dx_N = \int_I f\,.$$

Let us see the meaning of the given definition in the case $N = 3$. If $M = 1$, $\sigma : [a, b] \to \mathbb{R}^3$ is a curve and ω is a 1-differential form:

$$\omega(\boldsymbol{x}) = F_1(\boldsymbol{x})\, dx_1 + F_2(\boldsymbol{x})\, dx_2 + F_3(\boldsymbol{x})\, dx_3\,.$$

Hence,

$$\begin{aligned}\int_\sigma \omega &= \int_a^b [F_1(\sigma(t))\sigma_1'(t) + F_2(\sigma(t))\sigma_2'(t) + F_3(\sigma(t))\sigma_3'(t)]\, dt\\ &= \int_a^b \langle F(\sigma(t))\,, \sigma'(t)\rangle\, dt\,.\end{aligned}$$

This quantity will be called **line integral**[2] of the vector field $F = (F_1, F_2, F_3)$ along the curve σ, and will be denoted by

$$\int_\sigma \langle F, d\ell \rangle .$$

Example Let us compute the line integral of the vector field $F(x, y, z) = (-y, x, z^2)$ along the curve $\sigma : [0, 2\pi] \to \mathbb{R}^3$, defined by $\sigma(t) = (\cos t, \sin t, t)$:

$$\int_\sigma \langle F, d\ell \rangle = \int_0^{2\pi} [(\sin t)^2 + (\cos t)^2 + t^2]\, dt = 2\pi + \frac{8\pi^3}{3} .$$

If $M = 2$, $\sigma : [a_1, b_1] \times [a_2, b_2] \to \mathbb{R}^3$ is a surface and ω is a 2-differential form:

$$\omega(\mathbf{x}) = F_1(\mathbf{x})\, dx_2 \wedge dx_3 + F_2(\mathbf{x})\, dx_3 \wedge dx_1 + F_3(\mathbf{x})\, dx_1 \wedge dx_2 .$$

Hence,

$$\begin{aligned} \int_\sigma \omega = \int_{a_2}^{b_2} \int_{a_1}^{b_1} \Bigg[& F_1(\sigma(u, v)) \det \begin{pmatrix} \frac{\partial \sigma_2}{\partial u}(u, v) & \frac{\partial \sigma_2}{\partial v}(u, v) \\ \frac{\partial \sigma_3}{\partial u}(u, v) & \frac{\partial \sigma_3}{\partial v}(u, v) \end{pmatrix} + \\ & + F_2(\sigma(u, v)) \det \begin{pmatrix} \frac{\partial \sigma_3}{\partial u}(u, v) & \frac{\partial \sigma_3}{\partial v}(u, v) \\ \frac{\partial \sigma_1}{\partial u}(u, v) & \frac{\partial \sigma_1}{\partial v}(u, v) \end{pmatrix} + \\ & + F_3(\sigma(u, v)) \det \begin{pmatrix} \frac{\partial \sigma_1}{\partial u}(u, v) & \frac{\partial \sigma_1}{\partial v}(u, v) \\ \frac{\partial \sigma_2}{\partial u}(u, v) & \frac{\partial \sigma_2}{\partial v}(u, v) \end{pmatrix} \Bigg] du\, dv \\ = \int_{a_2}^{b_2} \int_{a_1}^{b_1} & \left\langle F(\sigma(u, v)), \frac{\partial \sigma}{\partial u}(u, v) \times \frac{\partial \sigma}{\partial v}(u, v) \right\rangle du\, dv . \end{aligned}$$

This quantity is called the **surface integral** or **flux**[3] of the vector field $F = (F_1, F_2, F_3)$ through the surface σ, and will be denoted by

$$\int_\sigma \langle F, dS \rangle .$$

[2]In mechanics this concept is used, for example, to define the **work** done by a field of forces on a particle moving along a curve.

[3]In fluidodynamics this concept is used, for instance, to define the amount of fluid crossing a given surface in the unit time.

Example Let us compute the flux of the vector field $F(x, y, z) = (-y, x, z^2)$ through the surface $\sigma : [0, 1] \times [0, 1] \to \mathbb{R}^3$, defined by $\sigma(u, v) = (u^2, v, u + v)$:

$$\int_\sigma \langle F, dS\rangle = \int_0^1 \int_0^1 [(-v)(-1) + u^2(-2u) + (u + v)^2(2u)]\, du\, dv = \frac{3}{2}.$$

3.8 Scalar Functions and M-Superficial Measure

We recall that, if ω is a M-differential form defined on a subset U of $\mathbb{R}^N$, with $1 \le M \le N$,

$$\omega(\boldsymbol{x}) = \sum_{1\le i_1<\cdots<i_M\le N} f_{i_1,\dots,i_M}(\boldsymbol{x})\, dx_{i_1} \wedge \cdots \wedge dx_{i_M},$$

and $\sigma : I \to \mathbb{R}^N$ is a M-surface whose support is contained in U, then

$$\int_\sigma \omega = \int_I \langle F(\sigma(\boldsymbol{u}))\,,\, \Sigma(\boldsymbol{u})\rangle\, d\boldsymbol{u},$$

where

$$F(\boldsymbol{x}) = \left(f_{i_1,\dots,i_M}(\boldsymbol{x})\right)_{1\le i_1<\cdots<i_M\le N},$$

$$\Sigma(\boldsymbol{u}) = \left(\det \sigma'_{(i_1,\dots,i_M)}(\boldsymbol{u})\right)_{1\le i_1<\cdots<i_M\le N}.$$

In view of the applications, besides the integral of a M-differential form, it is useful to define also the integral of a scalar function $f : U \to \mathbb{R}$ on a M-surface.

Definition 3.12

The function $f : U \to \mathbb{R}$ is **integrable** on the M-surface $\sigma : I \to \mathbb{R}^N$ if $(f \circ \sigma)\|\Sigma\|$ is integrable on I. In that case, we set

$$\begin{aligned}\int_\sigma f &= \int_I f(\sigma(\boldsymbol{u}))\, \|\Sigma(\boldsymbol{u})\|\, d\boldsymbol{u} \\ &= \int_I f(\sigma(\boldsymbol{u}))\left[\sum_{1\le i_1<\cdots<i_M\le N} \left(\det \sigma'_{(i_1,\dots,i_M)}(\boldsymbol{u})\right)^2\right]^{\frac{1}{2}} d\boldsymbol{u}.\end{aligned}$$

In this context, the integral does not differ for equivalent M-surfaces.

Theorem 3.13

If σ and $\tilde\sigma$ are two equivalent M-surfaces, then

$$\int_\sigma f = \int_{\tilde\sigma} f.$$

Proof

With the notations introduced previously, since $\sigma = \tilde{\sigma} \circ \varphi$, with $\varphi : A \to B$, we have

$$\begin{aligned} \Sigma(\boldsymbol{u}) &= \left(\det \sigma'_{(i_1,\dots,i_M)}(\boldsymbol{u})\right)_{1 \le i_1 < \dots < i_M \le N} \\ &= \left(\det \left(\tilde{\sigma}'_{(i_1,\dots,i_M)}(\varphi(\boldsymbol{u}))\varphi'(\boldsymbol{u})\right)\right)_{1 \le i_1 < \dots < i_M \le N} \\ &= \left(\det \tilde{\sigma}'_{(i_1,\dots,i_M)}(\varphi(\boldsymbol{u}))\right)_{1 \le i_1 < \dots < i_M \le N} \det \varphi'(\boldsymbol{u}) \\ &= \widetilde{\Sigma}(\varphi(\boldsymbol{u})) \det \varphi'(\boldsymbol{u}) \,. \end{aligned}$$

Therefore, by the Theorem on the Change of Variables in the integral, being $I \setminus A$ and $J \setminus B$ negligible, we have that

$$\begin{aligned} \int_\sigma f &= \int_A f(\sigma(\boldsymbol{u})) \, \|\Sigma(\boldsymbol{u})\| \, d\boldsymbol{u} \\ &= \int_A f(\tilde{\sigma}(\varphi(\boldsymbol{u}))) \, \|\widetilde{\Sigma}(\varphi(\boldsymbol{u}))\| \, |\det \varphi'(\boldsymbol{u})| \, d\boldsymbol{u} \\ &= \int_B f(\tilde{\sigma}(\boldsymbol{v})) \, \|\widetilde{\Sigma}(\boldsymbol{v})\| \, d\boldsymbol{v} \\ &= \int_{\tilde{\sigma}} f \,, \end{aligned}$$

thus proving the claim. □

In thc case $M = 1$, we have a curve $\sigma : [a, b] \to \mathbb{R}^N$ and, given a scalar function f defined on the support of σ,

$$\int_\sigma f = \int_a^b f(\sigma(t)) \, \|\sigma'(t)\| \, dt \,.$$

Consider the interesting case when f is constantly equal to 1 : having in mind the physical situation of a particle in motion along the curve described by σ, in this case the line integral is called the **length**[4] (or curvilinear measure) of the curve σ, and we write

$$\iota_1(\sigma) = \int_a^b \|\sigma'(t)\| \, dt \,.$$

[4]This definition could also be justified by geometrical considerations, which we omit here for the sake of briefness.

Example Let $\sigma : [0, b] \to \mathbb{R}^3$ be defined by $\sigma(t) = (t, t^2, 0)$. Its support is an arc of parabola, and its length is given by

$$\begin{aligned}\iota_1(\sigma) &= \int_0^b \sqrt{1+(2t)^2}\,dt \\ &= \int_{\sinh^{-1}(0)}^{\sinh^{-1}(2b)} \frac{1}{2}(\cosh u)^2\,du \\ &= \frac{1}{2}\left[\frac{u+\sinh u\cosh u}{2}\right]_0^{\sinh^{-1}(2b)} \\ &= \frac{1}{4}\left(\sinh^{-1}(2b) + 2b\sqrt{1+4b^2}\right) \\ &= \frac{1}{4}\ln\left(2b+\sqrt{1+4b^2}\right) + \frac{b}{2}\sqrt{1+4b^2}\,.\end{aligned}$$

If $M = 2$ and $N = 3$, we have the surface $\sigma : [a_1, b_1] \times [a_2, b_2] \to \mathbb{R}^3$ and, given a scalar function f, defined on the support of σ,

$$\int_\sigma f = \int_{a_2}^{b_2}\int_{a_1}^{b_1} f(\sigma(u,v))\left\|\frac{\partial\sigma}{\partial u}(u,v)\times\frac{\partial\sigma}{\partial v}(u,v)\right\|\,du\,dv\,.$$

Again it is interesting to consider the case when f is constantly equal to 1 : in this case we call **area** (or surface measure) of the surface σ the following integral:

$$\iota_2(\sigma) = \int_{a_2}^{b_2}\int_{a_1}^{b_1}\left\|\frac{\partial\sigma}{\partial u}(u,v)\times\frac{\partial\sigma}{\partial v}(u,v)\right\|\,du\,dv\,.$$

In the case when, for example, the surface happens to be a 2-parametrization of a set in $\mathbb{R}^3$, this integral is the flux of a vector field which at every point coincides with the normal versor to the surface itself.[5]

Example Let $\sigma : [0, \pi] \times [0, 2\pi] \to \mathbb{R}^3$ be defined by

$$\sigma(\phi, \theta) = (R\sin\phi\cos\theta,\ R\sin\phi\sin\theta,\ R\cos\phi)\,.$$

This is a 2-parametrization of a sphere of radius R, and its area is given by

$$\begin{aligned}\iota_2(\sigma) &= \int_0^{2\pi}\int_0^\pi \sqrt{(R^2\sin^2\phi\cos\theta)^2 + (R^2\sin^2\phi\sin\theta)^2 + (R^2\sin\phi\cos\theta)^2}\,d\phi\,d\theta \\ &= \int_0^{2\pi}\int_0^\pi R^2\sin\phi\,d\phi\,d\theta \\ &= 4\pi R^2\,.\end{aligned}$$

[5] Also the definition of the area of a surface can be justified by geometrical considerations, even if the procedure is much more delicate than in the case of a curve.

In general, the case when f is constantly equal to 1 gives

$$\int_\sigma 1 = \int_I \|\Sigma(\boldsymbol{u})\| \, d\boldsymbol{u} \,,$$

and leads to the following.

Definition 3.14

We call **M-superficial measure** of a M-surface $\sigma : I \to \mathbb{R}^N$ the following integral:

$$\iota_M(\sigma) = \int_I \left[\sum_{1 \le i_1 < \cdots < i_M \le N} \left(\det \sigma'_{(i_1,\dots,i_M)}(\boldsymbol{u}) \right)^2 \right]^{\frac{1}{2}} d\boldsymbol{u} \,.$$

As reasonably one expects, as a direct consequence of Theorem 3.13 and Theorem 3.8 one has the following.

Corollary 3.15

Two equivalent M-surfaces always have the same M-superficial measure. In particular, this is true for any two M-parametrizations of a given set.

Example Consider the two curves $\sigma, \tilde{\sigma} : [0, 2\pi] \to \mathbb{R}^2$, defined by

$$\sigma(t) = (\cos(t), \sin(t)) \,, \qquad \tilde{\sigma}(t) = (\cos(2t), \sin(2t)) \,.$$

Notice that, even if they have the same support, these curves are not equivalent. Indeed, as is easily seen, $\iota_1(\sigma) = 2\pi$ while $\iota_1(\tilde{\sigma}) = 4\pi$.

The above considerations naturally lead to the following.

Definition 3.16

We call **M-dimensional measure** of a M-parametrizable set $\mathcal{M} \subseteq \mathbb{R}^N$ the M-superficial measure of any of its M-parametrizations.

In the cases when $M = 1, 2$, the M-dimensional measure of $\mathcal{M}$ is often called **length** or **area** of $\mathcal{M}$, respectively. We may thus consider, for example, the length of a circle or the area of a sphere.[6]

If $M = N$, it can be verified that the N-dimensional measure of the set $\mathcal{M}$ is the same as the usual measure which has been treated in ▶ Chap. 2.

[6]We emphasize here the fact that the area of a sphere of radius R, which we found to be equal to $4\pi R^2$ by taking a particular parametrization, will always be the same when computed with *any* parametrization. This fact is not always proved in other textbooks.

Exercises

1. Find the length of the helicoidal curve $\gamma : [0, 2\pi] \to \mathbb{R}^3$ defined as

$$\gamma(t) = (\cos t, \sin t, t) .$$

2. Compute the integral

$$\int_\gamma xyz\, d\ell ,$$

where $\gamma : [0, 1] \to \mathbb{R}^3$ is defined as $\gamma(t) = (t, t^2, t^3)$.

3. Find the area of the ellipsoid

$$\left\{(x, y, z) \in \mathbb{R}^3 : \frac{x^2}{a^2} + \frac{y^2}{b^2} + \frac{z^2}{c^2} = 1\right\} .$$

4. Compute the integral

$$\int_\sigma (x + y + z)\, d\mathcal{S} ,$$

where $\sigma : [1, 2] \times [0, 1] \to \mathbb{R}^3$ is the surface defined as

$$\sigma(u, v) = (u \sin v, v \sin u, \cos(uv)) .$$

5. Find a parametrization $\sigma : I \to \mathbb{R}^3$ of the set

$$\mathcal{M} = \{(x, y, z) \in \mathbb{R}^3 : x^2 + 4y^2 + 9z^2 = 1\} ,$$

where $I \subseteq \mathbb{R}^2$ is some rectangle. Then, compute the integral $\int_\sigma f$, where $f : \mathbb{R}^3 \to \mathbb{R}$ is the function defined by the formula

$$f(x, y, z) = xyz .$$

6. Find a parametrization $\sigma : I \to \mathbb{R}^3$ of the set

$$\mathcal{M} = \{(x, y, z) \in \mathbb{R}^3 : |z| \leq 4x^2 + 9y^2 \leq 1\} ,$$

taking as $I \subseteq \mathbb{R}^3$ a rectangle (parallelepiped). Then, compute the volume of such a solid.

3.9 The Oriented Boundary of a Rectangle

Assume that $\sigma_1 : I_1 \to \mathbb{R}^N, \ldots, \sigma_n : I_n \to \mathbb{R}^N$ are some M-surfaces. We can easily find some equivalent M-surfaces $\tilde{\sigma}_1 : J_1 \to \mathbb{R}^N, \ldots, \tilde{\sigma}_n : J_n \to \mathbb{R}^N$, with the same orientation, such that the rectangles $J_1, \dots, J_n$ be non-overlapping and whose union be a rectangle I.

Definition 3.17

We call **glueing** of the M-surfaces $\sigma_1, \dots, \sigma_n$ any function $\sigma : I \to \mathbb{R}^N$ whose restrictions to $\mathring{J}_1, \dots, \mathring{J}_n$ coincide with $\tilde{\sigma}_1, \dots, \tilde{\sigma}_n$, respectively; it is almost everywhere differentiable, and we can define $\int_\sigma \omega$ by the same formula we have used for the M-surfaces of class $\mathcal{C}^1$. Hence,

$$\int_\sigma \omega = \int_{\sigma_1} \omega + \dots + \int_{\sigma_n} \omega .$$

We have thus "glued" together the M-surfaces $\sigma_1, \dots, \sigma_n$ and defined an integral of this glueing which does not depend on the particular choice of the equivalent M-surfaces, since anyway they conserve the orientation. In practice, however, we will never need to construct explicitly the glueing; what will actually be important is the formula for the integral.

Assume now that I be a rectangle[7] of $\mathbb{R}^{M+1}$, with $M \geq 1$:

$$I = [a_1, b_1] \times \dots \times [a_{M+1}, b_{M+1}] .$$

We denote by I_k the rectangle of $\mathbb{R}^M$ obtained from I by suppression of the k-th component:

$$I_k = [a_1, b_1] \times \dots \times [a_{k-1}, b_{k-1}] \times [a_{k+1}, b_{k+1}] \times \dots \times [a_{M+1}, b_{M+1}] .$$

Consider, for every k, the M-surfaces $\alpha_k^+, \beta_k^+ : I_k \to \mathbb{R}^{M+1}$ defined by

$$\alpha_k^+(u_1, \dots, \widehat{u_k}, .., u_{M+1}) = (u_1, \dots, u_{k-1}, a_k, u_{k+1}, \dots, u_{M+1}) ,$$

$$\beta_k^+(u_1, \dots, \widehat{u_k}, .., u_{M+1}) = (u_1, \dots, u_{k-1}, b_k, u_{k+1}, \dots, u_{M+1}) ,$$

where the meaning of the symbol $\widehat{\ }$ is to suppress the underlying variable. Consider moreover some M-surfaces $\alpha_k^-, \beta_k^- : I_k \to \mathbb{R}^{M+1}$, equivalent to α_k^+, β_k^+, respectively, with opposite orientation.

[7] We are thus considering here the situation when $N = M + 1$. This setting will be maintained also in the next section.

Definition 3.18
We call **oriented boundary** of the rectangle I a function ∂I which is a glueing of the following M-surfaces:

(a) α_k^- and β_k^+ if k is odd;

(b) α_k^+ and β_k^- if k is even.

If ω is a M-differential form defined on a subset U of $\mathbb{R}^{M+1}$ containing the image of ∂I, we will then have

$$\begin{aligned}\int_{\partial I} \omega &= \sum_{k=1}^{M+1} (-1)^k \int_{\alpha_k^+} \omega + \sum_{k=1}^{M+1} (-1)^{k-1} \int_{\beta_k^+} \omega \\ &= \sum_{k=1}^{M+1} (-1)^{k-1} \left(\int_{\beta_k^+} \omega - \int_{\alpha_k^+} \omega \right).\end{aligned}$$

Let $M = 1$, and consider the rectangle $[a_1, b_1] \times [a_2, b_2]$. Then, for example,

$$\begin{aligned}&\alpha_1^- : [a_2, b_2] \to \mathbb{R}^2, \quad v \mapsto (a_1, a_2 + b_2 - v),\\ &\beta_1^+ : [a_2, b_2] \to \mathbb{R}^2, \quad v \mapsto (b_1, v),\\ &\alpha_2^+ : [a_1, b_1] \to \mathbb{R}^2, \quad u \mapsto (u, a_2),\\ &\beta_2^- : [a_1, b_1] \to \mathbb{R}^2, \quad u \mapsto (a_1 + b_1 - u, b_2).\end{aligned}$$

We can visualize geometrically ∂I as the glueing of the sides of the rectangle I oriented in such a way that the perimeter be described in counter-clockwise direction.

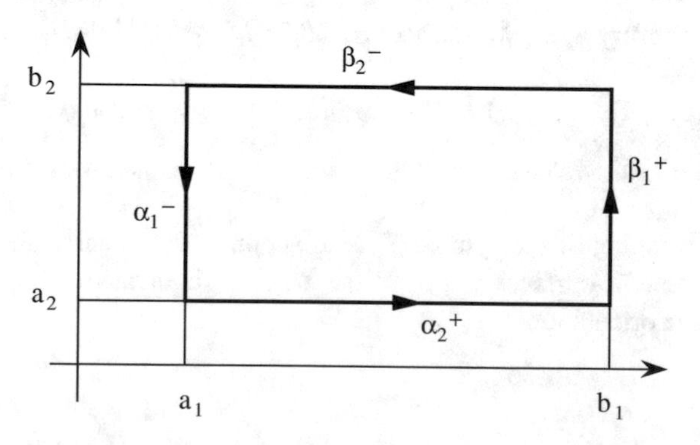

If $M = 2$, we have, for example,

$$\begin{aligned}&\alpha_1^- : [a_2, b_2] \times [a_3, b_3] \to \mathbb{R}^3, \quad (v, w) \mapsto (a_1, a_2 + b_2 - v, w),\\ &\beta_1^+ : [a_2, b_2] \times [a_3, b_3] \to \mathbb{R}^3, \quad (v, w) \mapsto (b_1, v, w),\\ &\alpha_2^+ : [a_1, b_1] \times [a_3, b_3] \to \mathbb{R}^3, \quad (u, w) \mapsto (u, a_2, w),\end{aligned}$$

$$\beta_2^- : [a_1, b_1] \times [a_3, b_3] \to \mathbb{R}^3, \quad (u, w) \mapsto (u, b_2, a_3 + b_3 - w),$$

$$\alpha_3^- : [a_1, b_1] \times [a_2, b_2] \to \mathbb{R}^3, \quad (u, v) \mapsto (a_1 + b_1 - u, v, a_3),$$

$$\beta_3^+ : [a_1, b_1] \times [a_2, b_2] \to \mathbb{R}^3, \quad (u, v) \mapsto (u, v, b_3).$$

In this case, we can visualize ∂I as the glueing of the six faces of the parallelepiped I, each oriented in such a way that the normal versor be always directed towards the exterior.

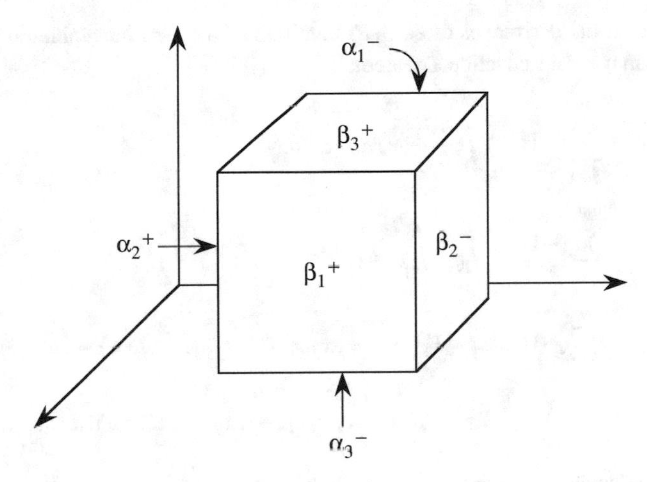

3.10 The Gauss Formula

In this section, I will be a rectangle in $\mathbb{R}^N$, with $N \geq 2$. In the following theorem, the elegant **Gauss formula** is obtained.[8]

Theorem 3.19
If ω is a $(N-1)$-differential form of class $\mathcal{C}^1$ defined on an open set containing the rectangle I in $\mathbb{R}^N$, then

$$\int_I d\omega = \int_{\partial I} \omega .$$

Proof
We can write ω as

$$\omega(\mathbf{r}) = \sum_{j=1}^{N} F_j(\mathbf{r})\, dx_1 \wedge \cdots \wedge \widehat{dx_j} \wedge \cdots \wedge dx_N ,$$

[8] As mentioned in the footnote in the previous section, we have $N = M + 1$.

Then,

$$d\omega(\boldsymbol{x}) = \sum_{j=1}^{N}\sum_{m=1}^{N} \frac{\partial F_j}{\partial x_m}(\boldsymbol{x})\, dx_m \wedge dx_1 \wedge \cdots \wedge \widehat{dx_j} \wedge \cdots \wedge dx_N$$

$$= \sum_{j=1}^{N}(-1)^{j-1} \frac{\partial F_j}{\partial x_j}(\boldsymbol{x})\, dx_1 \wedge \cdots \wedge dx_N\,.$$

Being the partial derivatives of each F_j continuous, they are integrable on the rectangle I, and we can use the Reduction Theorem:

$$\int_I d\omega = \sum_{j=1}^{N}(-1)^{j-1}\int_I \frac{\partial F_j}{\partial x_j}(\boldsymbol{x})\, dx_1 \dots dx_N$$

$$= \sum_{j=1}^{N}(-1)^{j-1}\int_{I_j}\left(\int_{a_j}^{b_j}\frac{\partial F_j}{\partial x_j}(x_1,\dots,x_N)\, dx_j\right) dx_1 \dots \widehat{dx_j} \dots dx_N$$

$$= \sum_{j=1}^{N}(-1)^{j-1}\int_{I_j}[F_j(x_1,\dots,x_{j-1},b_j,x_{j+1},\dots,x_N) -$$

$$-F_j(x_1,\dots,x_{j-1},a_j,x_{j+1},\dots,x_N)]\, dx_1 \dots \widehat{dx_j} \dots dx_N\,,$$

by the Fundamental Theorem. On the other hand, we have

$$\int_{\alpha_k^+}\omega = \sum_{j=1}^{N}\int_{\alpha_k^+} F_j\, dx_1 \wedge \cdots \wedge \widehat{dx_j} \wedge \cdots \wedge dx_N$$

$$= \sum_{j=1}^{N}\int_{I_k}(F_j \circ \alpha_k^+)\det(\alpha_k^+)'_{(1,\dots,\hat{j},\dots,N)}\, dx_1 \dots \widehat{dx_j} \dots dx_N$$

$$= \int_{I_k} F_k(x_1,\dots,x_{k-1},a_k,x_{k+1},\dots,x_N)\, dx_1 \dots \widehat{dx_k} \dots dx_N\,,$$

being

$$\det(\alpha_k^+)'_{(1,\dots,\hat{j},\dots,N)} = \begin{cases} 0 & \text{if } j \neq k\,,\\ 1 & \text{if } j = k\,.\end{cases}$$

Similarly,

$$\int_{\beta^+}\omega = \int_{I_k} F_k(x_1,\dots,x_{k-1},b_k,x_{k+1},\dots,x_N)\, dx_1 \dots \widehat{dx_k} \dots dx_N\,,$$

so that

$$\int_{\partial I} \omega = \sum_{k=1}^{N} (-1)^{k-1} \left(\int_{\beta_k^+} \omega - \int_{\alpha_k^+} \omega \right)$$

$$= \sum_{k=1}^{N} (-1)^{k-1} \int_{I_k} [F_k(x_1, \dots, x_{k-1}, b_k, x_{k+1}, \dots, x_N) -$$

$$- F_k(x_1, \dots, x_{k-1}, a_k, x_{k+1}, \dots, x_N)] \, dx_1 \dots \widehat{dx_k} \dots dx_N \,,$$

and the proof is completed. □

Remark The regularity assumption on the differential form ω could be considerably weakened. However, for briefness, we will not enter into this discussion. The interested reader is referred to [19].

3.11 Oriented Boundary of a M-Surface

In this section, I will be a rectangle in $\mathbb{R}^{M+1}$ and $\sigma : I \to \mathbb{R}^N$ a $(M + 1)$-surface.

Definition 3.20

For $1 \leq M \leq N - 1$, we call **oriented boundary** of σ a function $\partial\sigma = \sigma \circ \partial I$, which is a glueing of the following M-surfaces:

(a) $\sigma \circ \alpha_k^-$ and $\sigma \circ \beta_k^+$ if k is odd;

(b) $\sigma \circ \alpha_k^+$ and $\sigma \circ \beta_k^-$ if k is even.

Given a M-differential form ω whose domain contains the support of $\partial\sigma$, we will then have

$$\int_{\partial\sigma} \omega = \sum_{k=1}^{M+1} (-1)^k \int_{\sigma \circ \alpha_k^+} \omega + \sum_{k=1}^{M+1} (-1)^{k-1} \int_{\sigma \circ \beta_k^+} \omega$$

$$= \sum_{k=1}^{M+1} (-1)^{k-1} \left(\int_{\sigma \circ \beta_k^+} \omega - \int_{\sigma \circ \alpha_k^+} \omega \right).$$

Remark It is useful to extend the meaning of $\int_{\partial\sigma} \omega$ to the case when $\sigma : [a, b] \to \mathbb{R}^N$ is a curve, with $N \geq 1$, and $\omega = f : U \to \mathbb{R}$ is a 0-differential form; in this case, we set

$$\int_{\partial\sigma} \omega = f(\sigma(b)) - f(\sigma(a)).$$

Examples As an illustration, consider as usual the case $N = 3$. We begin with three examples of oriented boundaries of surfaces.

1. Let $\sigma : [r, R] \times [0, 2\pi] \to \mathbb{R}^3$, with $0 \le r < R$, be given by

$$\sigma(u, v) = (u \cos v, u \sin v, 0).$$

Its support is a disk if $r = 0$, an annulus if $r > 0$. The oriented boundary $\partial\sigma$ is given by a glueing of the following four curves:

$$\sigma \circ \alpha_1^-(v) = (r \cos v, -r \sin v, 0),$$
$$\sigma \circ \beta_1^+(v) = (R \cos v, R \sin v, 0),$$
$$\sigma \circ \alpha_2^+(u) = (u, 0, 0),$$
$$\sigma \circ \beta_2^-(u) = (r + R - u, 0, 0).$$

The first curve has as support a circle with radius r, which degenerates into the origin in the case when $r = 0$. The second has as support a circle with radius R. Notice however that these two circles are described by the two curves in opposite directions. The last two curves are equivalent with opposite orientations.

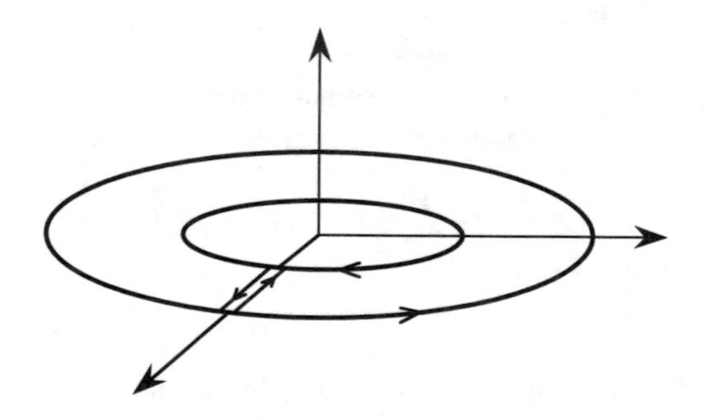

Consider, for example, the vector field $F(x, y, z) = (-y, x, xye^z)$. Then,

$$\begin{aligned}\int_{\partial\sigma} \langle F, d\ell\rangle &= \int_{\sigma\circ\alpha_1^-} \langle F, d\ell\rangle + \int_{\sigma\circ\beta_1^+} \langle F, d\ell\rangle \\ &= \int_0^{2\pi} [-r^2 \sin^2 v - r^2 \cos^2 v]\, dv + \int_0^{2\pi} [R^2 \sin^2 v + R^2 \cos^2 v]\, dv \\ &= 2\pi(R^2 - r^2).\end{aligned}$$

2. Consider the surface $\sigma : [r, R] \times [0, 2\pi] \to \mathbb{R}^3$, with $0 < r < R$, defined by

$$\sigma(u, v) = \left(\left(\frac{r+R}{2} + \left(u - \frac{r+R}{2} \right) \cos\left(\frac{v}{2} \right) \right) \cos v, \right.$$

$$\left(\frac{r+R}{2} + \left(u - \frac{r+R}{2} \right) \cos\left(\frac{v}{2} \right) \right) \sin v,$$

$$\left. \left(u - \frac{r+R}{2} \right) \sin\left(\frac{v}{2} \right) \right),$$

whose support is a Möbius strip. In this case, the oriented boundary is given by a glueing of

$$\sigma \circ \alpha_1^-(v) = \left(\left(\frac{r+R}{2} + \frac{R-r}{2} \cos\left(\frac{v}{2} \right) \right) \cos v, \right.$$

$$\left(\frac{r+R}{2} + \frac{R-r}{2} \cos\left(\frac{v}{2} \right) \right) \sin v,$$

$$\left. -\frac{R-r}{2} \sin\left(\frac{v}{2} \right) \right),$$

$$\sigma \circ \beta_1^+(v) = \left(\left(\frac{r+R}{2} + \frac{R-r}{2} \cos\left(\frac{v}{2} \right) \right) \cos v, \right.$$

$$\left(\frac{r+R}{2} + \frac{R-r}{2} \cos\left(\frac{v}{2} \right) \right) \sin v,$$

$$\left. \frac{R-r}{2} \sin\left(\frac{v}{2} \right) \right),$$

$$\sigma \circ \alpha_2^+(u) = (u, 0, 0),$$

$$\sigma \circ \beta_2^-(u) = (u, 0, 0).$$

Notice that in this case the two last curves are the same.

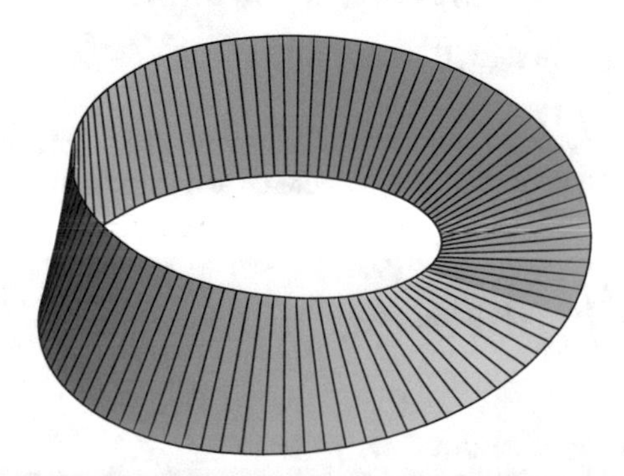

3. Consider the surface $\sigma : [0, \pi] \times [0, 2\pi] \to \mathbb{R}^3$ defined by

$$\sigma(\phi, \theta) = (R \sin\phi \cos\theta,\ R \sin\phi \sin\theta,\ R\cos\phi)\,,$$

whose support is the sphere with radius $R > 0$, centered at the origin. In this case, the oriented boundary is a glueing of

$$\sigma \circ \alpha_1^-(\theta) = (0, 0, R)\,,$$
$$\sigma \circ \beta_1^+(\theta) = (0, 0, -R)\,,$$
$$\sigma \circ \alpha_2^+(\phi) = (R\sin\phi, 0, R\cos\phi)\,,$$
$$\sigma \circ \beta_2^-(\phi) = (R\sin\phi, 0, -R\cos\phi)\,.$$

Notice that the first two curves are degenerated in one point, while the two others are equivalent with opposite orientations. Hence, for any choice of a vector field F, it will be $\int_{\partial\sigma} \langle F, d\ell\rangle = 0$.

Let us see now an example of oriented boundary of a volume in $\mathbb{R}^3$. Let $\sigma : [0, R] \times [0, \pi] \times [0, 2\pi] \to \mathbb{R}^3$ be the volume defined by

$$\sigma(\rho, \phi, \theta) = (\rho \sin\phi \cos\theta,\ \rho\sin\phi\sin\theta,\ \rho\cos\phi)\,,$$

whose support is the closed ball, centered at the origin, with radius $R > 0$. The oriented boundary $\partial\sigma$ is a glueing of the following six surfaces:

$$\sigma \circ \alpha_1^-(\phi, \theta) = (0, 0, 0)\,,$$
$$\sigma \circ \beta_1^+(\phi, \theta) = (R\sin\phi\cos\theta,\ R\sin\phi\sin\theta,\ R\cos\phi)\,,$$
$$\sigma \circ \alpha_2^+(\rho, \theta) = (0, 0, \rho)\,,$$
$$\sigma \circ \beta_2^-(\rho, \theta) = (0, 0, -\rho)\,,$$
$$\sigma \circ \alpha_3^-(\rho, \phi) = ((R-\rho)\sin\phi, 0, (R-\rho)\cos\phi)\,,$$
$$\sigma \circ \beta_3^+(\rho, \phi) = (\rho\sin\phi, 0, \rho\cos\phi)\,.$$

Notice that the first surface is degenerated in a point (the origin), the second has as support the entire sphere, the third and the fourth are degenerated in two lines, while the remaining two are equivalent with opposite orientations. Hence, given a vector field F, we will have

$$\int_{\partial\sigma} \langle F, dS\rangle = \int_{\sigma\circ\beta_1^+} \langle F, dS\rangle\,.$$

3.12 The Stokes–Cartan Formula

Let us state the following generalization of the Gauss theorem, where the important **Stokes–Cartan formula** is obtained.

Theorem 3.21
Let $0 \le M \le N-1$. If $\omega : U \to \Omega_M(\mathbb{R}^N)$ is a M-differential form of class $\mathcal{C}^1$ and $\sigma : I \to \mathbb{R}^N$ is a $(M+1)$-surface whose support is contained in U, then

$$\int_\sigma d\omega = \int_{\partial\sigma} \omega \,.$$

Notice that the case $M=0$, $N=1$ and $\sigma(u)=u$ is a version of the Fundamental Theorem, with the further assumption on the derivative of ω to be continuous.

The general proof of the above theorem is given in Appendix B. We will concentrate here on some corollaries obtained, in the case $N=3$, when M takes the values 0, 1 and 2. It is interesting to prove directly these corollaries, showing how the general proof adapts to them.

The Case $M=0$ We consider a 0-differential form $\omega = f : U \to \mathbb{R}$, and we obtain the following.

Theorem 3.22
Let $f : U \to \mathbb{R}$ be a scalar function of class $\mathcal{C}^1$ and $\sigma : [a,b] \to \mathbb{R}^3$ a curve whose support is contained in U. Then,

$$\int_\sigma \langle \text{grad}\, f, d\ell \rangle = f(\sigma(b)) - f(\sigma(a)) \,.$$

Proof
Consider the function $G : [a,b] \to \mathbb{R}$ defined by $G(t) = f(\sigma(t))$. It is of class $\mathcal{C}^1$, and by the Fundamental Theorem we have

$$\int_a^b G'(t)\, dt = G(b) - G(a) \,.$$

Since $G'(t) = \langle \text{grad}\, f(\sigma(t)), \sigma'(t) \rangle$, the conclusion follows. □

Remark The line integral of the gradient of a function f does not depend on the chosen curve itself, but only on the values of the function at the two extrema $\sigma(b)$ and $\sigma(a)$.

Example Given

$$F(x,y,z) = -\left(\frac{x}{[x^2+y^2+z^2]^{3/2}}, \frac{y}{[x^2+y^2+z^2]^{3/2}}, \frac{z}{[x^2+y^2+z^2]^{3/2}} \right)$$

and the curve $\sigma : [0, 4\pi] \to \mathbb{R}^3$ defined by $\sigma(t) = (\cos t, \sin t, t)$, we want to compute the line integral $\int_\sigma \langle F, d\ell \rangle$. Observe that $F = \text{grad } f$, with

$$f(x, y, z) = \frac{1}{\sqrt{x^2 + y^2 + z^2}}.$$

Hence,

$$\int_\sigma \langle F, d\ell \rangle = f(\sigma(4\pi)) - f(\sigma(0)) = \frac{1}{\sqrt{1 + 16\pi^2}} - 1.$$

The Case $M = 1$ We consider a 1-differential form

$$\omega(\boldsymbol{x}) = F_1(\boldsymbol{x})\, dx_1 + F_2(\boldsymbol{x})\, dx_2 + F_3(\boldsymbol{x})\, dx_3,$$

and we obtain the **Stokes–Ampère formula**.

Theorem 3.23
Let $F : U \to \mathbb{R}^3$ be a C^1-vector field and $\sigma : [a_1, b_1] \times [a_2, b_2] \to \mathbb{R}^3$ be a surface whose support is contained in U. Then,

$$\int_\sigma \langle \text{curl } F, dS \rangle = \int_{\partial\sigma} \langle F, d\ell \rangle.$$

Verbally *The flux of the curl of the vector field F through the surface σ is equal to the line integral of F along the oriented boundary of σ.*

Proof
Let $I = [a_1, b_1] \times [a_2, b_2]$, and define the following 1-differential form $\widetilde{\omega} : I \to \Omega_1(\mathbb{R}^2)$:

$$\widetilde{\omega}(u, v) = \left\langle F(\sigma(u, v)), \frac{\partial\sigma}{\partial u}(u, v) \right\rangle du + \left\langle F(\sigma(u, v)), \frac{\partial\sigma}{\partial v}(u, v) \right\rangle dv.$$

We first consider its integral on α_1^- :

$$\begin{aligned}\int_{\alpha_1^-} \widetilde{\omega} &= \int_{a_2}^{b_2} \left\langle F(\sigma(a_1, a_2 + b_2 - v)), -\frac{\partial\sigma}{\partial v}(a_1, a_2 + b_2 - v) \right\rangle dv \\ &= \int_{\sigma\circ\alpha_1^-} \langle F, d\ell \rangle.\end{aligned}$$

The integrals on β_1^+, α_2^+ and β_2^- are then treated analogously, so that

$$\int_{\partial I} \widetilde{\omega} = \int_{\partial\sigma} \langle F, d\ell \rangle.$$

Assume now that σ is of class $\mathcal{C}^2$. Then, $\widetilde{\omega}$ is of class $\mathcal{C}^1$ and, with some computations, we get

$$\begin{aligned} d\widetilde{\omega}(u,v) &= \left[\frac{\partial}{\partial u}\left\langle F(\sigma(u,v))\,,\,\frac{\partial\sigma}{\partial v}(u,v)\right\rangle - \frac{\partial}{\partial v}\left\langle F(\sigma(u,v))\,,\,\frac{\partial\sigma}{\partial u}(u,v)\right\rangle\right] du\wedge dv \\ &= \left\langle \operatorname{curl} F(\sigma(u,v))\,,\,\frac{\partial\sigma}{\partial u}(u,v)\times\frac{\partial\sigma}{\partial v}(u,v)\right\rangle du\wedge dv\,, \end{aligned}$$

so that

$$\int_I d\widetilde{\omega} = \int_\sigma \langle \operatorname{curl} F, d\mathcal{S}\rangle\,.$$

The Gauss formula applied to $\widetilde{\omega}$ thus yields the conclusion in this case.

The assumption that σ be of class $\mathcal{C}^2$ can eventually be eliminated by an approximation procedure: it is possible to construct a sequence $(\sigma_n)_n$ of surfaces of class $\mathcal{C}^2$ which converge to σ together with all their partial derivatives of the first order. The Stokes–Ampère formula holds then for those surfaces, and passing to the limit, by the Dominated Convergence Theorem, we have the conclusion. □

Example Let $F(x,y,z) = (-y,x,0)$ and $\gamma : [0,2\pi] \to \mathbb{R}^3$ be the curve defined by $\gamma(t) = (R\cos t, R\sin t, 0)$; we want to compute the line integral $\int_\gamma \langle F, d\ell\rangle$. We have already seen how to compute this integral by the direct use of the definition. We now proceed in a different way: consider the surface $\sigma : [0,R]\times[0,2\pi] \to \mathbb{R}^3$ given by $\sigma(\rho,\theta) = (\rho\cos\theta, \rho\sin\theta, 0)$. Observe that $\gamma = \sigma\circ\beta_1^+$, so

$$\int_\gamma \langle F, d\ell\rangle = \int_{\sigma\circ\beta_1^+} \langle F, d\ell\rangle = \int_{\partial\sigma} \langle F, d\ell\rangle = \int_\sigma \langle \operatorname{curl} F, d\mathcal{S}\rangle\,.$$

Being $\operatorname{curl} F(x,y,z) = (0,0,2)$ and

$$\frac{\partial\sigma}{\partial\rho}(\rho,\theta)\times\frac{\partial\sigma}{\partial\theta}(\rho,\theta) = (0,0,\rho)\,,$$

we then have

$$\int_\gamma \langle F, d\ell\rangle = \int_0^R\int_0^{2\pi} \langle (0,0,2)\,,\,(0,0,\rho)\rangle\, d\theta\, d\rho = 2\pi R^2\,.$$

The Case $M = 2$ We consider a 2-differential form

$$\omega(\boldsymbol{x}) = F_1(\boldsymbol{x})\,dx_2\wedge dx_3 + F_2(\boldsymbol{x})\,dx_3\wedge dx_1 + F_3(\boldsymbol{x})\,dx_1\wedge dx_2\,,$$

and we obtain the **Gauss–Ostrogradski formula**.

Theorem 3.24

Let $F : U \to \mathbb{R}^3$ be a C^1-vector field and $\sigma : I = [a_1, b_1] \times [a_2, b_2] \times [a_3, b_3] \to \mathbb{R}^3$ be a volume whose support is contained in U. Then,

$$\int_\sigma \operatorname{div} F \, dx_1 \wedge dx_2 \wedge dx_3 = \int_{\partial\sigma} \langle F, dS \rangle .$$

Hence, if σ is regular and injective on $\mathring{I}$, with $\det \sigma' > 0$, then

$$\int_{\sigma(I)} \operatorname{div} F = \int_{\partial\sigma} \langle F, dS \rangle .$$

In Intuitive Terms *The integral of the divergence of the vector field F on the set $V = \sigma(I)$ is equal to the flux of F which exits from V.*

Proof

Let $\widetilde{\omega} : I \to \Omega_2(\mathbb{R}^3)$ be the 2-differential form defined by

$$\begin{aligned}\widetilde{\omega}(\boldsymbol{u}) = & \left\langle F(\sigma(\boldsymbol{u})), \frac{\partial\sigma}{\partial u_2}(\boldsymbol{u}) \times \frac{\partial\sigma}{\partial u_3}(\boldsymbol{u}) \right\rangle du_2 \wedge du_3 + \\ & + \left\langle F(\sigma(\boldsymbol{u})), \frac{\partial\sigma}{\partial u_3}(\boldsymbol{u}) \times \frac{\partial\sigma}{\partial u_1}(\boldsymbol{u}) \right\rangle du_3 \wedge du_1 + \\ & + \left\langle F(\sigma(\boldsymbol{u})), \frac{\partial\sigma}{\partial u_1}(\boldsymbol{u}) \times \frac{\partial\sigma}{\partial u_2}(\boldsymbol{u}) \right\rangle du_1 \wedge du_2 .\end{aligned}$$

Considering β_1^+, we have:

$$\begin{aligned}\int_{\beta_1^+} \widetilde{\omega} &= \int_{a_2}^{b_2} \int_{a_3}^{b_3} \left\langle F(b_1, u_2, u_3), \frac{\partial\sigma}{\partial u_2}(b_1, u_2, u_3) \times \frac{\partial\sigma}{\partial u_3}(b_1, u_2, u_3) \right\rangle du_2 \, du_3 \\ &= \int_{\beta_1^+} \langle F, dS \rangle .\end{aligned}$$

With the analogous computations on the remaining five surfaces which determine ∂I we can say that

$$\int_{\partial I} \widetilde{\omega} = \int_{\partial\sigma} \langle F, dS \rangle .$$

Assume now that σ be of class $\mathcal{C}^2$. Then $\widetilde{\omega}$ is of class $\mathcal{C}^1$ and, carrying over the computations, with some tenacity, we have

$$\begin{aligned} d\widetilde{\omega}(\boldsymbol{u}) &= \left[\frac{\partial}{\partial u_1}\left\langle F(\sigma(\boldsymbol{u}))\,,\, \frac{\partial\sigma}{\partial u_2}(\boldsymbol{u}) \times \frac{\partial\sigma}{\partial u_3}(\boldsymbol{u})\right\rangle + \right. \\ &\quad + \frac{\partial}{\partial u_2}\left\langle F(\sigma(\boldsymbol{u}))\,,\, \frac{\partial\sigma}{\partial u_3}(\boldsymbol{u}) \times \frac{\partial\sigma}{\partial u_1}(\boldsymbol{u})\right\rangle + \\ &\quad \left. + \frac{\partial}{\partial u_3}\left\langle F(\sigma(\boldsymbol{u}))\,,\, \frac{\partial\sigma}{\partial u_1}(\boldsymbol{u}) \times \frac{\partial\sigma}{\partial u_2}(\boldsymbol{u})\right\rangle\right] du_1 \wedge du_2 \wedge du_3 \\ &= \operatorname{div} F(\sigma(\boldsymbol{u})) \det \sigma'(\boldsymbol{u})\, du_1 \wedge du_2 \wedge du_3\,. \end{aligned}$$

Hence,

$$\int_I d\widetilde{\omega} = \int_I \operatorname{div} F(\sigma(\boldsymbol{u})) \det \sigma'(\boldsymbol{u})\, du_1 \wedge du_2 \wedge du_3 = \int_\sigma \operatorname{div} F\, dx_1 \wedge dx_2 \wedge dx_3.$$

The Gauss formula applied to $\widetilde{\omega}$ thus yields the first conclusion. On the other hand, if σ is regular and injective on $\mathring{I}$, with $\det \sigma' > 0$, Theorem 3.11 yields the second formula, and the proof is completed. The assumption that σ be of class $\mathcal{C}^2$ can eventually be eliminated, as in the previous theorem. □

Example We want to compute the flux of the vector field

$$F(x, y, z) = ([x^2 + y^2 + z^2]x, [x^2 + y^2 + z^2]y, [x^2 + y^2 + z^2]z)$$

through a spherical surface parametrized by $\eta : [0, \pi] \times [0, 2\pi] \to \mathbb{R}^3$, defined as

$$\eta(\phi, \theta) = (R \sin\phi \cos\theta, R \sin\phi \sin\theta, R \cos\phi)\,.$$

Recall that $\eta = \sigma \circ \beta_1^+$, where $\sigma : I = [0, R] \times [0, \pi] \times [0, 2\pi] \to \mathbb{R}^3$ is the volume given by

$$\sigma(\rho, \phi, \theta) = (\rho \sin\phi \cos\theta, \rho \sin\phi \sin\theta, \rho \cos\phi)\,.$$

Hence,

$$\int_\eta \langle F, dS\rangle = \int_{\partial\sigma} \langle F, dS\rangle = \int_{\sigma(I)} \operatorname{div} F\,.$$

Being $\operatorname{div} F(x, y, z) = 5(x^2 + y^2 + z^2)$, passing to spherical coordinates we have

$$\int_{\sigma(I)} \operatorname{div} F = \int_0^{2\pi}\int_0^{\pi}\int_0^{R} (5\rho^2)(\rho^2 \sin\phi)\, d\rho\, d\phi\, d\theta = 4\pi R^5\,.$$

3.13 Analogous Results in $\mathbb{R}^2$

Assuming that U be a subset of $\mathbb{R}^2$ we find two interesting corollaries of the Stokes–Cartan theorem. Like in the case $N = 3$, the line integral of a vector field $F = (F_1, F_2)$ along a curve $\sigma : [a, b] \to \mathbb{R}^2$ is defined as

$$\int_\sigma \langle F, d\ell \rangle = \int_a^b [F_1(\sigma(t))\sigma_1'(t) + F_2(\sigma(t))\sigma_2'(t)]\, dt\,.$$

We have the following result, analogous to the one obtained in the previous section for the case $N = 3$.

Theorem 3.25
Let $f : U \to \mathbb{R}$ be a scalar function of class $\mathcal{C}^1$ and $\sigma : [a, b] \to \mathbb{R}^2$ be a curve with support contained in U. Then,

$$\int_\sigma \left\langle \left(\frac{\partial f}{\partial x_1}, \frac{\partial f}{\partial x_2} \right), d\ell \right\rangle = f(\sigma(b)) - f(\sigma(a))\,.$$

Taking now $M = 1$, we obtain the **Gauss–Green formula.**

Theorem 3.26
Let $F = (F_1, F_2) : U \to \mathbb{R}^2$ be a $\mathcal{C}^1$-vector field and $\sigma : I = [a_1, b_1] \times [a_2, b_2] \to \mathbb{R}^2$ be a surface whose support is contained in U. Then,

$$\int_\sigma \left(\frac{\partial F_2}{\partial x_1} - \frac{\partial F_1}{\partial x_2} \right) dx_1 \wedge dx_2 = \int_{\partial\sigma} \langle F, d\ell \rangle\,.$$

Hence, if σ is regular and injective on $\mathring{I}$, with $\det \sigma' > 0$, *then*

$$\int_{\sigma(I)} \left(\frac{\partial F_2}{\partial x_1} - \frac{\partial F_1}{\partial x_2} \right) = \int_{\partial\sigma} \langle F, d\ell \rangle\,.$$

Proof
As for the Stokes–Ampère theorem, we consider the auxiliary differential form $\widetilde{\omega} : I \to \Omega_1(\mathbb{R}^2)$, defined by

$$\widetilde{\omega}(u, v) = \left\langle F(\sigma(u, v)), \frac{\partial \sigma}{\partial u}(u, v) \right\rangle du + \left\langle F(\sigma(u, v)), \frac{\partial \sigma}{\partial v}(u, v) \right\rangle dv\,,$$

and see that

$$\int_{\partial I} \widetilde{\omega} = \int_{\partial \sigma} \langle F, d\ell \rangle .$$

If σ is of class $\mathcal{C}^2$, then $\widetilde{\omega}$ is of class $\mathcal{C}^1$ and computation shows that

$$\begin{aligned} d\widetilde{\omega}(u,v) &= \left[\frac{\partial}{\partial u} \left\langle F(\sigma(u,v)) , \frac{\partial \sigma}{\partial v}(u,v) \right\rangle - \frac{\partial}{\partial v} \left\langle F(\sigma(u,v)) , \frac{\partial \sigma}{\partial u}(u,v) \right\rangle \right] du \wedge dv \\ &= \left(\frac{\partial F_2}{\partial x_1}(\sigma(u,v)) - \frac{\partial F_1}{\partial x_2}(\sigma(u,v)) \right) \det \sigma'(u,v)\, du \wedge dv . \end{aligned}$$

Hence,

$$\begin{aligned} \int_I d\widetilde{\omega} &= \int_I \left(\frac{\partial F_2}{\partial x_1}(\sigma(u,v)) - \frac{\partial F_1}{\partial x_2}(\sigma(u,v)) \right) \det \sigma'(u,v)\, du \wedge dv \\ &= \int_\sigma \left(\frac{\partial F_2}{\partial x_1} - \frac{\partial F_1}{\partial x_2} \right) dx_1 \wedge dx_2 . \end{aligned}$$

The Gauss formula applied to $\widetilde{\omega}$ gives us the first conclusion. If σ is regular and injective on $\mathring{I}$, with $\det \sigma' > 0$, then Theorem 3.11 yields the conclusion. As mentioned in the case $N = 3$, the assumption that σ is of class $\mathcal{C}^2$ can be eliminated by an approximation procedure. □

Example Consider the surface $\sigma : I = [0,1] \times [0, 2\pi] \to \mathbb{R}^2$ defined by $\sigma(\rho, \theta) = (A\rho \cos\theta, B\rho \sin\theta)$, whose support is an elliptical surface with semi-axes having lengths $A > 0$ and $B > 0$. Take the vector field $F(x,y) = (-y, x)$. Being

$$\frac{\partial F_2}{\partial x}(x,y) - \frac{\partial F_1}{\partial y}(x,y) = 2 ,$$

and (as for the disk)

$$\int_{\partial \sigma} \langle F, d\ell \rangle = \int_{\sigma \circ \beta_1^+} \langle F, d\ell \rangle ,$$

the Gauss–Green formula gives us

$$\int_{\sigma(I)} 2\, dx\, dy = \int_0^{2\pi} \langle (-B \sin\theta, A \cos\theta) , (-A \sin\theta, B \cos\theta) \rangle \, d\theta = 2\pi AB .$$

We then find the area of the elliptic surface: $\mu(\sigma(I)) = \pi AB$.

Exercises

1. Let $\gamma : [0, 4] \to \mathbb{R}^2$ be the curve defined as $\gamma(t) = (t, \sqrt{t})$. Compute the integral

$$\int_\gamma \langle (xy^2, yx^2), d\ell \rangle ,$$

first directly, then by the use of a scalar potential.

2. Let $\sigma : [0, 1] \times [0, 1] \to \mathbb{R}^2$ be the surface defined by

$$\sigma(u, v) = (u^2, v^2) .$$

Compute $\int_{\partial\sigma} \langle F, d\ell \rangle$ with $F : \mathbb{R}^2 \to \mathbb{R}^2$ given by

$$F(x, y) = (x^2 - y^2, 2xy) .$$

3. Let $\sigma : [0, 1] \times [0, 1] \to \mathbb{R}^3$ be the surface defined by

$$\sigma(u, v) = (u^2, v^2, uv) .$$

Compute $\int_\sigma \langle \mathrm{curl} F, dS \rangle$, both directly and by the use of the Stokes–Ampère formula, when

$$F(x, y, z) = (x - y, y - z, z - x) .$$

4. Let $\sigma : [0, 1] \times [0, 1] \times [-1, 1] \to \mathbb{R}^3$ be the volume defined as

$$\sigma(u, v, w) = (u^2 + v^2, u^2 + v^2, w) ,$$

and consider the vector field $F : \mathbb{R}^3 \to \mathbb{R}^3$ given by

$$F(x, y, z) = (x^2, y^3, z^4) .$$

Compute $\int_\sigma \mathrm{div} F \, dx \wedge dy \wedge dz$, both directly and by the use of the Gauss–Ostrogradski formula.

5. Let $\sigma : [0, 1] \times [0, \pi] \times [0, 2\pi] \to \mathbb{R}^3$ be the parametrization of the unit ball in spherical coordinates, given by

$$\sigma(\rho, \phi, \theta) = (\rho \sin\phi \cos\theta, \rho \sin\phi \sin\theta, \rho \cos\phi) .$$

Compute $\int_{\partial\sigma} \langle F, dS \rangle$, where $F : \mathbb{R}^3 \to \mathbb{R}^3$ is the vector field defined as

$$F(x, y, z) = (x - y + z , y - z + x , z - x + y) .$$

3.14 Exact Differential Forms

We are now interested in the following problem. Given a differential form ω, when is it possible to write it as the external differential of another differential form $\widetilde{\omega}$, to be determined? In the following, we assume $M \geq 1$.

Definition 3.27
A M-differential form ω is said to be **closed** if $d\omega = 0$; it is said to be **exact** if there is a $(M-1)$-differential form $\widetilde{\omega}$ such that $d\widetilde{\omega} = \omega$.

Every exact differential form is closed: if $\omega = d\widetilde{\omega}$, then $d\omega = d(d\widetilde{\omega}) = 0$. The contrary is not always true.

Example The 1-differential form defined on $\mathbb{R}^2 \setminus \{(0,0)\}$ by

$$\omega(x, y) = \frac{-y}{x^2 + y^2}\, dx + \frac{x}{x^2 + y^2}\, dy$$

is closed, as easily verified: setting

$$F_1(x, y) = \frac{-y}{x^2 + y^2}\,, \qquad F_2(x, y) = \frac{x}{x^2 + y^2}\,,$$

for every $(x, y) \neq (0, 0)$, it is

$$\frac{\partial F_2}{\partial x}(x, y) = \frac{\partial F_1}{\partial y}(x, y)\,.$$

Let us compute the line integral of its vector field $F = (F_1, F_2)$ on the curve $\sigma : [0, 2\pi] \to \mathbb{R}^2$, defined by $\sigma(t) = (\cos t, \sin t)$:

$$\begin{aligned}
\int_\sigma \langle F, d\ell \rangle &= \int_0^{2\pi} \langle F(\sigma(t)), \sigma'(t) \rangle \, dt \\
&= \int_0^{2\pi} \langle (-\sin t, \cos t), (-\sin t, \cos t) \rangle \, dt \\
&= 2\pi\,.
\end{aligned}$$

Assume by contradiction that ω be exact, i.e., that there exists a C^1-function $f : \mathbb{R}^2 \setminus \{(0,0)\} \to \mathbb{R}$ such that $\frac{\partial f}{\partial x} = F_1$ and $\frac{\partial f}{\partial y} = F_2$. In that case, being $\sigma(0) = \sigma(2\pi)$, we should have

$$\int_\sigma \langle F, d\ell \rangle = \int_\sigma \left\langle \left(\frac{\partial f}{\partial x}, \frac{\partial f}{\partial y} \right), d\ell \right\rangle = f(\sigma(2\pi)) - f(\sigma(0)) = 0\,,$$

which contradicts the above.

The **Poincaré theorem** says that the situation of the preceding example can never happen if, for example, the open set U on which the differential form is defined is **star-shaped**.

Definition 3.28

A set U is star-shaped with respect to a point $\bar{x}$ if, with each of its points x, the set U contains the whole segment joining x to $\bar{x}$, i.e.,

$$\{\bar{x} + t(x - \bar{x}) : t \in [0, 1]\} \subseteq U .$$

For example, every convex set is star-shaped (with respect to any of its points). In particular, a ball, or a rectangle, or even the whole space $\mathbb{R}^N$ are star-shaped. Clearly, the set $\mathbb{R}^2 \setminus \{(0, 0)\}$ considered above is not star-shaped.

Theorem 3.29

Let U be an open subset of $\mathbb{R}^N$, star-shaped with respect to a point $\bar{x}$. For $1 \le M \le N$, a M-differential form $\omega : U \to \Omega_M(\mathbb{R}^N)$ of class $\mathcal{C}^1$ is exact if and only if it is closed. In that case, if ω is of the type

$$\omega(x) = \sum_{1 \le i_1 < \cdots < i_M \le N} f_{i_1,\dots,i_M}(x)\, dx_{i_1} \wedge \cdots \wedge dx_{i_M} ,$$

a $(M - 1)$-differential form $\widetilde{\omega}$ such that $d\widetilde{\omega} = \omega$ is given by

$$\widetilde{\omega}(x) = \sum_{1 \le i_1 < \cdots < i_M \le N} \sum_{s=1}^{M} (-1)^{s+1} (x_{i_s} - \bar{x}_{i_s}) \cdot$$

$$\cdot \left(\int_0^1 t^{M-1} f_{i_1,\dots,i_M}(\bar{x} + t(x - \bar{x}))\, dt \right) dx_{i_1} \wedge \cdots \wedge \widehat{dx_{i_s}} \wedge \cdots \wedge dx_{i_M} .$$

The proof of this theorem will be given in Appendix B. As for the Stokes–Cartan theorem, we consider here some corollaries which hold true for the case $N = 3$, giving for each of them a direct proof. In order to simplify the notations, we will assume without loss of generality that $\bar{x} = (0, 0, 0)$.

The Case $M = 1$ A $\mathcal{C}^1$-vector field $F = (F_1, F_2, F_3)$, defined on an open subset U of $\mathbb{R}^3$, determines a 1-differential form

$$\omega(x) = F_1(x)\, dx_1 + F_2(x)\, dx_2 + F_3(x)\, dx_3 .$$

This is closed if and only if $\operatorname{curl} F = 0$. In this case, the vector field is said to be **irrotational**. On the other hand, the vector field is said to be F **conservative** if there is a function $f : U \to \mathbb{R}$ such that $F = \operatorname{grad} f$. In that case, f is a **scalar potential** of the vector field F.[9]

Theorem 3.30
If $U \subseteq \mathbb{R}^3$ is star-shaped with respect to the origin, then the vector field $F : U \to \mathbb{R}^3$ is conservative if and only if it is irrotational, and in that case a function $f : U \to \mathbb{R}$ such that $F = \operatorname{grad} f$ is given by

$$f(\boldsymbol{x}) = \int_0^1 \langle F(t\boldsymbol{x}), \boldsymbol{x}\rangle \, dt \,.$$

Any other function $\tilde{f} : U \to \mathbb{R}$ which is such that $F = \operatorname{grad} \tilde{f}$ is obtained from f by adding a constant.

Proof
Set $\widetilde{\omega} = f : U \to \mathbb{R}$. Let us verify that $d\widetilde{\omega} = \omega$. Using the fact that $\operatorname{curl} F = 0$ and the Leibniz rule, we have

$$\begin{aligned}
\frac{\partial \widetilde{\omega}}{\partial x_j}(\boldsymbol{x}) &= \int_0^1 \frac{\partial}{\partial x_j} \langle F(t\boldsymbol{x}), \boldsymbol{x}\rangle \, dt \\
&= \int_0^1 \left(\sum_{i=1}^{3} \left(\frac{\partial F_i}{\partial x_j}(t\boldsymbol{x}) t x_i \right) + F_j(t\boldsymbol{x}) \right) dt \\
&= \int_0^1 \left(\sum_{i=1}^{3} \left(\frac{\partial F_j}{\partial x_i}(t\boldsymbol{x}) t x_i \right) + F_j(t\boldsymbol{x}) \right) dt \,.
\end{aligned}$$

Defining $\phi(t) = t F_j(t\boldsymbol{x})$, since

$$\phi'(t) = \sum_{i=1}^{3} \left(\frac{\partial F_j}{\partial x_i}(t\boldsymbol{x}) t x_i \right) + F_j(t\boldsymbol{x}) \,,$$

by the Fundamental Theorem we have

$$\frac{\partial \widetilde{\omega}}{\partial x_j}(\boldsymbol{x}) = \Big[t F_j(t\boldsymbol{x}) \Big]_{t=0}^{t=1} = F_j(\boldsymbol{x}) \,.$$

This proves that $F = \operatorname{grad} f$. The second part of the theorem follows directly from the fact that, if $\operatorname{grad} f = \operatorname{grad} \tilde{f}$, then $f - \tilde{f}$ has to be constant on U. □

[9]Beware that in Mechanics it is often the function $-f$ which is called "the potential".

Example Consider the vector field $F(x, y, z) = (2xz + y, x, x^2)$ which, as easily verified, is irrotational. A scalar potential is then given by

$$f(x, y, z) = \int_0^1 ((2t^2x^2z + txy) + txy + t^2x^2z)\, dt = xy + x^2z\,.$$

The Case $M = 2$ A $\mathcal{C}^1$-vector field $F = (F_1, F_2, F_3)$, defined on an open subset U of $\mathbb{R}^3$, determines a 2-differential form

$$\omega(\boldsymbol{x}) = F_1(\boldsymbol{x})\, dx_2 \wedge dx_3 + F_2(\boldsymbol{x})\, dx_3 \wedge dx_1 + F_3(\boldsymbol{x})\, dx_1 \wedge dx_2\,.$$

This is closed if and only if $\operatorname{div} F = 0$. In this case, the vector field is said to be **solenoidal**. One says that F has a **vector potential** if there is a vector field $V = (V_1, V_2, V_3)$ such that $F = \operatorname{curl} V$.

Theorem 3.31

If $U \subseteq \mathbb{R}^3$ is star-shaped with respect to the origin, then the vector field $F : U \to \mathbb{R}^3$ has a vector potential if and only if it is solenoidal, and in that case a vector field $V : U \to \mathbb{R}^3$ for which $F = \operatorname{curl} V$ is given by

$$\begin{aligned} V(\boldsymbol{x}) = \Big(&\int_0^1 t(F_2(t\boldsymbol{x})x_3 - F_3(t\boldsymbol{x})x_2)\, dt\,, \\ &\int_0^1 t(F_3(t\boldsymbol{x})x_1 - F_1(t\boldsymbol{x})x_3)\, dt\,, \\ &\int_0^1 t(F_1(t\boldsymbol{x})x_2 - F_2(t\boldsymbol{x})x_1)\, dt \Big)\,, \end{aligned}$$

which we will briefly write as

$$V(\boldsymbol{x}) = \int_0^1 t(F(t\boldsymbol{x}) \times \boldsymbol{x})\, dt\,.$$

Any other vector field $\widetilde{V} : U \to \mathbb{R}^3$ such that $F = \operatorname{curl} \widetilde{V}$ is obtained from V by adding the gradient of an arbitrary scalar function.

Proof

Consider the 1-differential form determined by the vector field V,

$$\begin{aligned} \widetilde{\omega}(\boldsymbol{x}) = &\left(\int_0^1 t(F_2(t\boldsymbol{x})x_3 - F_3(t\boldsymbol{x})x_2)\, dt \right) dx_1 + \\ &+ \left(\int_0^1 t(F_3(t\boldsymbol{x})x_1 - F_1(t\boldsymbol{x})x_3)\, dt \right) dx_2 + \\ &+ \left(\int_0^1 t(F_1(t\boldsymbol{x})x_2 - F_2(t\boldsymbol{x})x_1)\, dt \right) dx_3\,. \end{aligned}$$

We have to prove that $d\tilde{\omega} = \omega$. By the Leibniz rule, taking into account the fact that ω is closed, we find

$$\begin{aligned}
&\frac{\partial}{\partial x_2}\int_0^1 t(F_1(t\boldsymbol{x})x_2 - F_2(t\boldsymbol{x})x_1)\,dt - \frac{\partial}{\partial x_3}\int_0^1 t(F_3(t\boldsymbol{x})x_1 - F_1(t\boldsymbol{x})x_3)\,dt = \\
&\quad = \int_0^1 \left(t^2\left(\frac{\partial F_1}{\partial x_1}(t\boldsymbol{x})x_1 + \frac{\partial F_1}{\partial x_2}(t\boldsymbol{x})x_2 + \frac{\partial F_1}{\partial x_3}(t\boldsymbol{x})x_3\right) + 2tF_1(t\boldsymbol{x})\right)dt \\
&\quad = F_1(\boldsymbol{x}),
\end{aligned}$$

by applying the Fundamental Theorem to the function $\phi(t) = t^2F_1(t\boldsymbol{x})$. Analogously one obtains the remaining two identities, thus proving the formula. The second part of the theorem follows from the fact that, if curl $V =$ curl $\widetilde{V}$, then, by the previous theorem, $V - \widetilde{V}$ is a conservative vector field. □

Example Consider the solenoidal vector field $F(x, y, z) = (y, z, x)$. A vector potential is then given by

$$V(x, y, z) = \int_0^1 t(ty, tz, tx) \times (x, y, z)\,dt = \frac{1}{3}(z^2 - xy, x^2 - yz, y^2 - xz)\,.$$

The Case $M = 3$ A C^1-scalar function f, defined on an open subset U of $\mathbb{R}^3$, determines a 3-differential form

$$\omega(\boldsymbol{x}) = f(\boldsymbol{x})\,dx_1 \wedge dx_2 \wedge dx_3\,.$$

This is necessarily always closed, since $d\omega$ is a 4-differential form defined on a subset of $\mathbb{R}^3$.

Theorem 3.32
If $U \subseteq \mathbb{R}^3$ is star-shaped with respect to the origin, the function $f : U \to \mathbb{R}$ is always of the type $f =$ div W, where $W : U \to \mathbb{R}^3$ is the vector field defined by

$$W(\boldsymbol{x}) = \left(\int_0^1 t^2 f(t\boldsymbol{x})\,dt\right)\boldsymbol{x}\,.$$

Any other vector field $\tilde{W} : U \to \mathbb{R}^3$ such that $F =$ div $\tilde{W}$ is obtained from W by adding the curl of an arbitrary vector field.

Proof
Using the Leibniz rule, we have

$$\frac{\partial}{\partial x_1}\int_0^1 t^2 f(t\boldsymbol{x})x_1\,dt + \frac{\partial}{\partial x_2}\int_0^1 t^2 f(t\boldsymbol{x})x_2\,dt + \frac{\partial}{\partial x_3}\int_0^1 t^2 f(t\boldsymbol{x})x_3\,dt$$
$$= \int_0^1 \left(t^3\left(\frac{\partial f}{\partial x_1}(t\boldsymbol{x}) + \frac{\partial f}{\partial x_2}(t\boldsymbol{x}) + \frac{\partial f}{\partial x_3}(t\boldsymbol{x})\right) + 3t^2 f(t\boldsymbol{x})\right) dt$$
$$= f(\boldsymbol{x}),$$

by applying the Fundamental Theorem to the function $\phi(t) = t^3 f(t\boldsymbol{x})$. The second part of the theorem follows from the fact that, if div W = div $\tilde{W}$, then, by the preceding theorem, $W - \tilde{W}$ has a vector potential. □

Example Consider the scalar function $f(x, y, z) = xyz$. Then, a vector field whose divergence is f is given by

$$W(x, y, z) = \left(\int_0^1 t^5 xyz\,dt\right)(x, y, z) = \frac{1}{6}(x^2yz, xy^2z, xyz^2).$$

Exercises

1. Let $F : \mathbb{R}^3 \to \mathbb{R}^3$ be the vector field defined by the formula

$$F(x, y, z) = (e^x yz + e^y z + e^z y\,,\; e^x z + e^y xz + e^z x\,,\; e^x y + e^y x + e^z xy),$$

and let ω be the associated 1-differential form. Prove that ω is exact, and find a 0-differential form $\tilde{\omega}$ such that $d\tilde{\omega} = \omega$. Compute then the integral

$$\int_\gamma \omega\,,$$

where $\gamma : [0, 1] \to \mathbb{R}^3$ is the curve defined as

$$\gamma(t) = (t, t^2, t^3).$$

2. Let $F : \mathbb{R}^3 \setminus \{z = 0\} \to \mathbb{R}^3$ be the vector field defined as

$$F(x, y, z) = \left(\frac{2xy^2}{z^2}, \frac{2x^2y}{z^2}, \frac{-2x^2y^2}{z^3}\right).$$

Prove that it is conservative, and find a scalar potential.

3. Let $F : \mathbb{R}^3 \to \mathbb{R}^3$ be defined as

$$F(x, y, z) = \left(-2xyz\,, \frac{xz}{2}\,, yz^2\right).$$

Prove that it is solenoidal, and find a vector potential. Finally, compute the flux $\int_\sigma F \cdot d\mathcal{S}$, where $\sigma : [0, 2\pi] \times [0, \pi] \to \mathbb{R}^3$ is defined by

$$\sigma(u, v) = (\cos u \sin v\,,\ 2 \sin u \sin v\,,\ 3 \cos v)\,.$$

4. Let $\omega : \mathbb{R}^4 \to \Omega_3(\mathbb{R}^4)$ be the 3-differential form defined by

$$\begin{aligned}\omega(x_1, x_2, x_3, x_4) = \; & x_1 x_2\, dx_2 \wedge dx_3 \wedge dx_4 + x_2^2\, dx_1 \wedge dx_3 \wedge dx_4 + \\ & + x_2 x_3\, dx_1 \wedge dx_2 \wedge dx_4 + x_1^2 x_2\, dx_1 \wedge dx_2 \wedge dx_3\,.\end{aligned}$$

Prove that ω is exact, and find a 2-differential form $\tilde{\omega}$ such that $d\tilde{\omega} = \omega$.

5. Prove that the differential form

$$\omega : \mathbb{R}^3 \setminus (\{x = 0\} \cup \{y = 0\} \cup \{z = 0\}) \to \Omega_1(\mathbb{R}^3)\,,$$

defined as

$$\omega(x, y, z) = \frac{1}{x^2 yz}\, dx + \frac{1}{xy^2 z}\, dy + \frac{1}{xyz^2}\, dz,$$

is exact, and find a 0-differential form $\tilde{\omega}$ such that $d\tilde{\omega} = \omega$.

Appendix A
Differential Calculus in $\mathbb{R}^N$

Alessandro Fonda

A. Fonda, *The Kurzweil-Henstock Integral for Undergraduates*,
Compact Textbooks in Mathematics,
https://doi.org/10.1007/978-3-319-95321-2

Let $\Omega \subset \mathbb{R}^N$ be an open set, $\boldsymbol{x}_0$ a point of Ω, and $f : \Omega \to \mathbb{R}^M$ be a given function. We want to extend the notion of derivative already known in the case $M = N = 1$. Here is the definition.

Definition A.1

We say that f is differentiable at $\boldsymbol{x}_0$ if there exists a linear function $\ell : \mathbb{R}^N \to \mathbb{R}^M$ for which one can write

$$f(\boldsymbol{x}) = f(\boldsymbol{x}_0) + \ell(\boldsymbol{x} - \boldsymbol{x}_0) + r(\boldsymbol{x}),$$

where r is a function satisfying

$$\lim_{\boldsymbol{x} \to \boldsymbol{x}_0} \frac{r(\boldsymbol{x})}{\|\boldsymbol{x} - \boldsymbol{x}_0\|} = 0.$$

If f is differentiable at $\boldsymbol{x}_0$, the linear function ℓ is called **differential** of f at $\boldsymbol{x}_0$, and is denoted by

$$df(\boldsymbol{x}_0).$$

Following the tradition for linear functions, taking $\boldsymbol{h} \in \mathbb{R}^N$ we will often write $df(\boldsymbol{x}_0)\boldsymbol{h}$ instead of $df(\boldsymbol{x}_0)(\boldsymbol{h})$.

We will now review the main results needed in this book concerning the differential calculus.

A.1 The Differential of a Scalar-Valued Function

Assume first, for simplicity, that $M = 1$. We start by fixing a **direction** i.e., a vector $\boldsymbol{v} \in \mathbb{R}^N$ with $\|\boldsymbol{v}\| = 1$, also called a **versor**. Whenever it exists, we call **directional derivative** of f at $\boldsymbol{x}_0$ in the direction $\boldsymbol{v}$ the limit

$$\lim_{t\to 0} \frac{f(\boldsymbol{x}_0 + t\boldsymbol{v}) - f(\boldsymbol{x}_0)}{t},$$

which will be denoted by

$$\frac{\partial f}{\partial \boldsymbol{v}}(\boldsymbol{x}_0).$$

If $\boldsymbol{v}$ coincides with an element $\boldsymbol{e}_k$ of the canonical basis $(\boldsymbol{e}_1, \boldsymbol{e}_2, \dots, \boldsymbol{e}_N)$ of $\mathbb{R}^N$, the directional derivative is called k-th **partial derivative** of f at $\boldsymbol{x}_0$ and is denoted by

$$\frac{\partial f}{\partial x_k}(\boldsymbol{x}_0).$$

If $\boldsymbol{x}_0 = (x_1^0, x_2^0, \dots, x_N^0)$, then

$$\begin{aligned}\frac{\partial f}{\partial x_k}(\boldsymbol{x}_0) &= \lim_{t\to 0} \frac{f(\boldsymbol{x}_0 + t\boldsymbol{e}_k) - f(\boldsymbol{x}_0)}{t} \\ &= \lim_{t\to 0} \frac{f(x_1^0, x_2^0, \dots, x_k^0 + t, \dots, x_N^0) - f(x_1^0, x_2^0, \dots, x_k^0, \dots, x_N^0)}{t},\end{aligned}$$

so that it is frequent to call it "partial derivative with respect to the k-th variable".

Theorem A.2

If f is differentiable at $\boldsymbol{x}_0$, then f is continuous at $\boldsymbol{x}_0$. Moreover, all the directional derivatives of f at $\boldsymbol{x}_0$ exist: for every direction $\boldsymbol{v} \in \mathbb{R}^N$ one has

$$\frac{\partial f}{\partial \boldsymbol{v}}(\boldsymbol{x}_0) = df(\boldsymbol{x}_0)\boldsymbol{v}.$$

Proof

We know that the function $\ell = df(\boldsymbol{x}_0)$, being linear, is continuous, and $\ell(\boldsymbol{0}) = 0$. Then,

$$\begin{aligned}\lim_{\boldsymbol{x}\to\boldsymbol{x}_0} f(\boldsymbol{x}) &= \lim_{\boldsymbol{x}\to\boldsymbol{x}_0} [f(\boldsymbol{x}_0) + \ell(\boldsymbol{x} - \boldsymbol{x}_0) + r(\boldsymbol{x})] \\ &= f(\boldsymbol{x}_0) + \ell(\boldsymbol{0}) + \lim_{\boldsymbol{x}\to\boldsymbol{x}_0} r(\boldsymbol{x})\end{aligned}$$

$$
\begin{aligned}
&= f(\boldsymbol{x}_0) + \lim_{\boldsymbol{x}\to\boldsymbol{x}_0} \frac{r(\boldsymbol{x})}{\|\boldsymbol{x}-\boldsymbol{x}_0\|} \lim_{\boldsymbol{x}\to\boldsymbol{x}_0} \|\mathbf{r}-\mathbf{r}_0\| \\
&= f(\boldsymbol{x}_0)\,,
\end{aligned}
$$

showing that f is continuous at $\boldsymbol{x}_0$. Concerning the directional derivatives, we have

$$
\begin{aligned}
\lim_{t\to 0} \frac{f(\boldsymbol{x}_0+t\boldsymbol{v}) - f(\boldsymbol{x}_0)}{t} &= \lim_{t\to 0} \frac{df(\boldsymbol{x}_0)(t\boldsymbol{v}) + r(\boldsymbol{x}_0+t\boldsymbol{v})}{t} \\
&= \lim_{t\to 0} \frac{t\, df(\boldsymbol{x}_0)\boldsymbol{v} + r(\boldsymbol{x}_0+t\boldsymbol{v})}{t} \\
&= df(\boldsymbol{x}_0)\boldsymbol{v} + \lim_{t\to 0} \frac{r(\boldsymbol{x}_0+t\boldsymbol{v})}{t}\,.
\end{aligned}
$$

On the other hand, being $\|\boldsymbol{v}\| = 1$, it is

$$
\lim_{t\to 0} \left| \frac{r(\boldsymbol{x}_0+t\boldsymbol{v})}{t} \right| = \lim_{\boldsymbol{x}\to\boldsymbol{x}_0} \frac{|r(\boldsymbol{x})|}{\|\boldsymbol{x}-\boldsymbol{x}_0\|} = 0\,,
$$

whence the conclusion. □

In particular, if $\boldsymbol{v}$ coincides with an element $\boldsymbol{e}_k$ of the canonical basis $(\boldsymbol{e}_1, \boldsymbol{e}_2, \dots, \boldsymbol{e}_N)$, then

$$
\frac{\partial f}{\partial x_k}(\boldsymbol{x}_0) = df(\boldsymbol{x}_0)\boldsymbol{e}_k\,.
$$

Writing the vector $\boldsymbol{h} \in \mathbb{R}^N$ as $\boldsymbol{h} = h_1\boldsymbol{e}_1 + h_2\boldsymbol{e}_2 + \cdots + h_N\boldsymbol{e}_N$, by linearity we have

$$
\begin{aligned}
df(\boldsymbol{x}_0)\boldsymbol{h} &= h_1 df(\boldsymbol{x}_0)\boldsymbol{e}_1 + h_2 df(\boldsymbol{x}_0)\boldsymbol{e}_2 + \cdots + h_N df(\boldsymbol{x}_0)\boldsymbol{e}_N \\
&= h_1 \frac{\partial f}{\partial x_1}(\boldsymbol{x}_0) + h_2 \frac{\partial f}{\partial x_2}(\boldsymbol{x}_0) + \cdots + h_N \frac{\partial f}{\partial x_N}(\boldsymbol{x}_0)\,,
\end{aligned}
$$

i.e.,

$$
df(\boldsymbol{x}_0)\boldsymbol{h} = \sum_{k=1}^{N} \frac{\partial f}{\partial x_k}(\boldsymbol{x}_0) h_k\,.
$$

Theorem A.3

If f has partial derivatives defined in a neighborhood of $\boldsymbol{x}_0$, and they are continuous at $\boldsymbol{x}_0$, then f is differentiable at $\boldsymbol{x}_0$.

Proof

In order to simplify the notations, we will assume that $N = 2$. We define the function $\ell : \mathbb{R}^2 \to \mathbb{R}$ associating to every vector $\boldsymbol{h} = (h_1, h_2)$ the real number

$$\ell(\boldsymbol{h}) = \frac{\partial f}{\partial x_1}(\boldsymbol{x}_0)h_1 + \frac{\partial f}{\partial x_2}(\boldsymbol{x}_0)h_2 .$$

We will prove that ℓ is indeed the differential of f at $\boldsymbol{x}_0$. First of all, it is readily verified that it is linear. Moreover, writing $\boldsymbol{x}_0 = (x_1^0, x_2^0)$ and $\boldsymbol{x} = (x_1, x_2)$, by the Lagrange Mean Value Theorem one has

$$\begin{aligned} f(\boldsymbol{x}) - f(\boldsymbol{x}_0) &= (f(x_1, x_2) - f(x_1^0, x_2)) + (f(x_1^0, x_2) - f(x_1^0, x_2^0)) \\ &= \frac{\partial f}{\partial x_1}(\xi_1, x_2)(x_1 - x_1^0) + \frac{\partial f}{\partial x_2}(x_1^0, \xi_2)(x_2 - x_2^0), \end{aligned}$$

for some $\xi_1 \in]x_1^0, x_1[$ and $\xi_2 \in]x_2^0, x_2[$. Hence,

$$\begin{aligned} r(\boldsymbol{x}) &= f(\boldsymbol{x}) - f(\boldsymbol{x}_0) - \ell(\boldsymbol{x} - \boldsymbol{x}_0) \\ &= \left[\frac{\partial f}{\partial x_1}(\xi_1, x_2) - \frac{\partial f}{\partial x_1}(x_1^0, x_2^0)\right](x_1 - x_1^0) + \left[\frac{\partial f}{\partial x_2}(x_1^0, \xi_2) - \frac{\partial f}{\partial x_2}(x_1^0, x_2^0)\right](x_2 - x_2^0). \end{aligned}$$

Then, being $|x_1 - x_1^0| \le \|\boldsymbol{x} - \boldsymbol{x}_0\|$ and $|x_2 - x_2^0| \le \|\boldsymbol{x} - \boldsymbol{x}_0\|$,

$$\frac{|r(\boldsymbol{x})|}{\|\boldsymbol{x} - \boldsymbol{x}_0\|} \le \left|\frac{\partial f}{\partial x_1}(\xi_1, x_2) - \frac{\partial f}{\partial x_1}(x_1^0, x_2^0)\right| + \left|\frac{\partial f}{\partial x_2}(x_1^0, \xi_2) - \frac{\partial f}{\partial x_2}(x_1^0, x_2^0)\right| .$$

Letting $\boldsymbol{x}$ tend to $\boldsymbol{x}_0$, we have that $(\xi_1, x_2) \to (x_1^0, x_2^0)$ and $(x_1^0, \xi_2) \to (x_1^0, x_2^0)$ so that, being $\frac{\partial f}{\partial x_1}$ and $\frac{\partial f}{\partial x_2}$ continuous at $\boldsymbol{x}_0 = (x_1^0, x_2^0)$, it has to be

$$\lim_{\boldsymbol{x} \to \boldsymbol{x}_0} \frac{|r(\boldsymbol{x})|}{\|\boldsymbol{x} - \boldsymbol{x}_0\|} = 0,$$

whence the conclusion. □

We say that $f : \Omega \to \mathbb{R}$ is of class $\mathcal{C}^1$, or a $\mathcal{C}^1$-function, if f has partial derivatives which are continuous on the whole domain Ω. From the previous theorem we have that a function of class $\mathcal{C}^1$ is **differentiable**, i.e., differentiable at every point of Ω.

A.2 Twice Differentiable Scalar-Valued Functions

Let Ω be an open subset of $\mathbb{R}^N$, and $f : \Omega \to \mathbb{R}$ be a differentiable function. We want to extend the notion of "second derivative", which is well-known in the case $N = 1$. For simplicity, let us deal with the case $N = 2$. If the partial derivatives $\frac{\partial f}{\partial x_1}, \frac{\partial f}{\partial x_2} : \Omega \to \mathbb{R}$

have themselves the partial derivatives at a point $\boldsymbol{x}_0$, these are said to be "second order partial derivatives" of f at $\boldsymbol{x}_0$ and are denoted by

$$\frac{\partial^2 f}{\partial x_1^2}(\boldsymbol{x}_0) = \frac{\partial}{\partial x_1}\frac{\partial f}{\partial x_1}(\boldsymbol{x}_0), \qquad \frac{\partial^2 f}{\partial x_2 \partial x_1}(\boldsymbol{x}_0) = \frac{\partial}{\partial x_2}\frac{\partial f}{\partial x_1}(\boldsymbol{x}_0),$$

$$\frac{\partial^2 f}{\partial x_1 \partial x_2}(\boldsymbol{x}_0) = \frac{\partial}{\partial x_1}\frac{\partial f}{\partial x_2}(\boldsymbol{x}_0), \qquad \frac{\partial^2 f}{\partial x_2^2}(\boldsymbol{x}_0) = \frac{\partial}{\partial x_2}\frac{\partial f}{\partial x_2}(\boldsymbol{x}_0).$$

We now prove the **Schwarz Theorem.**

Theorem A.4

If the second order partial derivatives $\frac{\partial^2 f}{\partial x_2 \partial x_1}, \frac{\partial^2 f}{\partial x_1 \partial x_2}$ *exist in a neighborhood of* $\boldsymbol{x}_0$ *and they are continuous at* $\boldsymbol{x}_0$, *then*

$$\frac{\partial^2 f}{\partial x_2 \partial x_1}(\boldsymbol{x}_0) = \frac{\partial^2 f}{\partial x_1 \partial x_2}(\boldsymbol{x}_0).$$

Proof

Let $\rho > 0$ be such that $B(\boldsymbol{x}_0, \rho) \subseteq \Omega$.[1] We write $\boldsymbol{x}_0 = (x_1^0, x_2^0)$ and we take an $\boldsymbol{x} = (x_1, x_2) \in B(\boldsymbol{x}_0, \rho)$ such that $x_1 \neq x_1^0$ and $x_2 \neq x_2^0$. It is then possible to define

$$g(x_1, x_2) = \frac{f(x_1, x_2) - f(x_1, x_2^0)}{x_2 - x_2^0}, \qquad h(x_1, x_2) = \frac{f(x_1, x_2) - f(x_1^0, x_2)}{x_1 - x_1^0}.$$

One can verify that

$$\frac{g(x_1, x_2) - g(x_1^0, x_2)}{x_1 - x_1^0} = \frac{h(x_1, x_2) - h(x_1, x_2^0)}{x_2 - x_2^0}.$$

By the Lagrange Mean Value Theorem, there is a $\xi_1 \in]x_1^0, x_1[$ such that

$$\frac{g(x_1, x_2) - g(x_1^0, x_2)}{x_1 - x_1^0} = \frac{\partial g}{\partial x_1}(\xi_1, x_2) = \frac{\frac{\partial f}{\partial x_1}(\xi_1, x_2) - \frac{\partial f}{\partial x_1}(\xi_1, x_2^0)}{x_2 - x_2^0},$$

and there is a $\xi_2 \in]x_2^0, x_2[$ such that

$$\frac{h(x_1, x_2) - h(x_1, x_2^0)}{x_2 - x_2^0} = \frac{\partial h}{\partial x_2}(x_1, \xi_2) = \frac{\frac{\partial f}{\partial x_2}(x_1, \xi_2) - \frac{\partial f}{\partial x_2}(x_1^0, \xi_2)}{x_1 - x_1^0}.$$

[1] Let us recall the notation $B(\boldsymbol{x}_0, \rho)$ for the open ball centered at $\boldsymbol{x}_0$ with radius $\rho > 0$, and $\overline{B}(\boldsymbol{x}_0, \rho)$ for the closed ball.

Again by the Lagrange Mean Value Theorem, there is a $\eta_2 \in]x_2^0, x_2[$ such that

$$\frac{\frac{\partial f}{\partial x_1}(\xi_1, x_2) - \frac{\partial f}{\partial x_1}(\xi_1, x_2^0)}{x_2 - x_2^0} = \frac{\partial^2 f}{\partial x_2 \partial x_1}(\xi_1, \eta_2),$$

and there is a $\eta_1 \in]x_1^0, x_1[$ such that

$$\frac{\frac{\partial f}{\partial x_2}(x_1, \xi_2) - \frac{\partial f}{\partial x_2}(x_1^0, \xi_2)}{x_1 - x_1^0} = \frac{\partial^2 f}{\partial x_1 \partial x_2}(\eta_1, \xi_2).$$

Hence,

$$\frac{\partial^2 f}{\partial x_2 \partial x_1}(\xi_1, \eta_2) = \frac{\partial^2 f}{\partial x_1 \partial x_2}(\eta_1, \xi_2).$$

Taking the limit, as $\boldsymbol{x} = (x_1, x_2)$ tends to $\boldsymbol{x}_0 = (x_1^0, x_2^0)$, we have that both (ξ_1, η_2) and (η_1, ξ_2) converge to $\boldsymbol{x}_0$, and the continuity of the second order partial derivatives leads to the conclusion. □

We say that $f : \Omega \to \mathbb{R}$ is of class $\mathcal{C}^2$, or a $\mathcal{C}^2$-function, if all its second order partial derivatives exist and are continuous on Ω.

It could be useful to consider the **Hessian matrix** of f at $\boldsymbol{x}_0$:

$$Hf(\boldsymbol{x}_0) = \begin{pmatrix} \frac{\partial^2 f}{\partial x_1^2}(\boldsymbol{x}_0) & \frac{\partial^2 f}{\partial x_2 \partial x_1}(\boldsymbol{x}_0) \\ \\ \frac{\partial^2 f}{\partial x_1 \partial x_2}(\boldsymbol{x}_0) & \frac{\partial^2 f}{\partial x_2^2}(\boldsymbol{x}_0) \end{pmatrix};$$

if f is of class $\mathcal{C}^2$, this is a symmetric matrix.

What has been said above extends without difficulties for any $N \geq 2$. If f is of class $\mathcal{C}^2$, the Hessian matrix is an $N \times N$ symmetric matrix.

One can further define by induction the n–th order partial derivatives. It is said that $f : \Omega \to \mathbb{R}$ is of class $\mathcal{C}^n$ if all its n–th order partial derivatives exist and are continuous on Ω.

A.3 The Differential of a Vector-Valued Function

When $M \geq 2$, let $f_1, f_2, \dots, f_M$ be the components of the function $f : \Omega \to \mathbb{R}^M$, so that

$$f(\boldsymbol{x}) = (f_1(\boldsymbol{x}), f_2(\boldsymbol{x}), \dots, f_M(\boldsymbol{x}))\,.$$

Theorem A.5
The function f is differentiable at $\boldsymbol{x}_0$ if and only if such are all its components. In this case, for any vector $\boldsymbol{h} \in \mathbb{R}^N$ it is

$$df(\boldsymbol{x}_0)\boldsymbol{h} = (df_1(\boldsymbol{x}_0)\boldsymbol{h}, df_2(\boldsymbol{x}_0)\boldsymbol{h}, \dots, df_M(\boldsymbol{x}_0)\boldsymbol{h}).$$

Proof
Considering the components in the equation

$$f(\boldsymbol{x}) = f(\boldsymbol{x}_0) + \ell(\boldsymbol{x} - \boldsymbol{x}_0) + r(\boldsymbol{x}),$$

we can write

$$f_j(\boldsymbol{x}) = f_j(\boldsymbol{x}_0) + \ell_j(\boldsymbol{x} - \boldsymbol{x}_0) + r_j(\boldsymbol{x}),$$

with $j = 1, 2, \dots, M$, and we know that

$$\lim_{\boldsymbol{x}\to\boldsymbol{x}_0} \frac{r(\boldsymbol{x})}{\|\boldsymbol{x} - \boldsymbol{x}_0\|} = \boldsymbol{0} \quad \Longleftrightarrow \quad \lim_{\boldsymbol{x}\to\boldsymbol{x}_0} \frac{r_j(\boldsymbol{x})}{\|\boldsymbol{x} - \boldsymbol{x}_0\|} = 0 \quad \text{for every } j = 1, 2, \dots, M,$$

whence the conclusion. □

It is useful to consider the matrix associated to the linear function $\ell = df(\boldsymbol{x}_0)$, given by

$$\begin{pmatrix} \ell_1(\boldsymbol{e}_1) & \ell_1(\boldsymbol{e}_2) & \dots & \ell_1(\boldsymbol{e}_N) \\ \ell_2(\boldsymbol{e}_1) & \ell_2(\boldsymbol{e}_2) & \dots & \ell_2(\boldsymbol{e}_N) \\ \vdots & \vdots & & \vdots \\ \ell_M(\boldsymbol{e}_1) & \ell_M(\boldsymbol{e}_2) & \dots & \ell_M(\boldsymbol{e}_N) \end{pmatrix},$$

where $\boldsymbol{e}_1, \boldsymbol{e}_2, \dots \boldsymbol{e}_N$ are the vectors of the canonical basis of $\mathbb{R}^N$. This matrix is called **Jacobian matrix** associated to the function f at $\boldsymbol{x}_0$, and is denoted by $Jf(\boldsymbol{x}_0)$. Recalling that

$$\frac{\partial f_j}{\partial x_k}(\boldsymbol{x}_0) = df_j(\boldsymbol{x}_0)\boldsymbol{e}_k,$$

with $j = 1, 2, \dots, M$ and $k = 1, 2, \dots, N$, we obtain the matrix

$$Jf(\boldsymbol{x}_0) = \begin{pmatrix} \frac{\partial f_1}{\partial x_1}(\boldsymbol{x}_0) & \frac{\partial f_1}{\partial x_2}(\boldsymbol{x}_0) & \cdots & \frac{\partial f_1}{\partial x_N}(\boldsymbol{x}_0) \\ \frac{\partial f_2}{\partial x_1}(\boldsymbol{x}_0) & \frac{\partial f_2}{\partial x_2}(\boldsymbol{x}_0) & \cdots & \frac{\partial f_2}{\partial x_N}(\boldsymbol{x}_0) \\ \vdots & \vdots & & \vdots \\ \frac{\partial f_M}{\partial x_1}(\boldsymbol{x}_0) & \frac{\partial f_M}{\partial x_2}(\boldsymbol{x}_0) & \cdots & \frac{\partial f_M}{\partial x_N}(\boldsymbol{x}_0) \end{pmatrix}.$$

The function $f : \Omega \to \mathbb{R}^M$ is said to be of class $\mathcal{C}^1$, or $\mathcal{C}^2$, if all its components are such. This definition naturally extends to functions of class $\mathcal{C}^n$.

A.4 Some Computational Rules

Let us start with some easy propositions.

1. If $f : \Omega \to Y$ is constant, then $df(x_0) = 0$, for every $x_0 \in \Omega$.
2. If $\mathcal{A} : \Omega \to Y$ is linear and continuous, then $d\mathcal{A}(x_0) = \mathcal{A}$, for every $x_0 \in \Omega$.
3. If $\mathbb{R}^N = \mathbb{R}^{N_1} \times \mathbb{R}^{N_2}$ and $\mathcal{B} : \Omega \to Y$ is bilinear and continuous, writing $x_0 = (x_1^0, x_2^0)$ and $h = (h_1, h_2)$, with $x_1^0, h_1 \in \mathbb{R}^{N_1}$ and $x_2^0, h_2 \in \mathbb{R}^{N_2}$, one has

$$d\mathcal{B}(x_0)(h) = \mathcal{B}(x_1^0, h_2) + \mathcal{B}(h_1, x_2^0)\,.$$

All this can be generalized to n-linear continuous functions.

We now recall the usual laws of calculus.

Theorem A.6
If $f, g : \Omega \to Y$ are differentiable at x_0 and α, β are two real numbers, then

$$d(\alpha f + \beta g)(x_0) = \alpha df(x_0) + \beta dg(x_0)\,.$$

Proof
Writing

$$f(x) = f(x_0) + df(x_0)(x - x_0) + r_1(x)\,, \quad g(x) = g(x_0) + dg(x_0)(x - x_0) + r_2(x)\,,$$

we have that

$$(\alpha f + \beta g)(x) = (\alpha f + \beta g)(x_0) + (\alpha df(x_0) + \beta dg(x_0))(x - x_0) + r(x)\,,$$

with $r(x) = \alpha r_1(x) + \beta r_2(x)$, and

$$\lim_{x \to x_0} \frac{r(x)}{\|x - x_0\|} = \alpha \lim_{x \to x_0} \frac{r_1(x)}{\|x - x_0\|} + \beta \lim_{x \to x_0} \frac{r_2(x)}{\|x - x_0\|} = 0\,.$$

Hence, $\alpha f + \beta g$ is differentiable at x_0 with differential $\alpha df(x_0) + \beta dg(x_0)$. □

We now study the differentiability of a composite function.

Theorem A.7
If $f : \Omega \to \mathbb{R}^M$ is differentiable at $\boldsymbol{x}_0$, while $\Omega' \subseteq \mathbb{R}^M$ is an open set containing $f(\Omega)$ and $g : \Omega' \to \mathbb{R}^L$ is differentiable at $f(\boldsymbol{x}_0)$, then $g \circ f$ is differentiable at $\boldsymbol{x}_0$, and

$$d(g \circ f)(\boldsymbol{x}_0) = dg(f(\boldsymbol{x}_0)) \circ df(\boldsymbol{x}_0)\,.$$

Proof
Setting $\boldsymbol{y}_0 = f(\boldsymbol{x}_0)$, we have

$$f(\boldsymbol{x}) = f(\boldsymbol{x}_0) + df(\boldsymbol{x}_0)(\boldsymbol{x} - \boldsymbol{x}_0) + r_1(\boldsymbol{x})\,, \quad g(\boldsymbol{y}) = g(\boldsymbol{y}_0) + dg(\boldsymbol{y}_0)(\boldsymbol{y} - \boldsymbol{y}_0) + r_2(\boldsymbol{y})\,,$$

with

$$\lim_{\boldsymbol{x} \to \boldsymbol{x}_0} \frac{r_1(\boldsymbol{x})}{\|\boldsymbol{x} - \boldsymbol{x}_0\|} = \boldsymbol{0}\,, \qquad \lim_{\boldsymbol{y} \to \boldsymbol{y}_0} \frac{r_2(\boldsymbol{y})}{\|\boldsymbol{y} - \boldsymbol{y}_0\|} = \boldsymbol{0}\,.$$

Let us introduce the function $R_2 : \Omega' \to \mathbb{R}^L$, defined as

$$R_2(\boldsymbol{y}) = \begin{cases} \dfrac{r_2(\boldsymbol{y})}{\|\boldsymbol{y} - \boldsymbol{y}_0\|} & \text{if } \boldsymbol{y} \neq \boldsymbol{y}_0\,, \\ \boldsymbol{0} & \text{if } \boldsymbol{y} = \boldsymbol{y}_0\,. \end{cases}$$

Notice that R_2 is continuous at $\boldsymbol{y}_0$. Then,

$$\begin{aligned} g(f(\boldsymbol{x})) &= g(f(\boldsymbol{x}_0)) + dg(f(\boldsymbol{x}_0))[f(\boldsymbol{x}) - f(\boldsymbol{x}_0)] + r_2(f(\boldsymbol{x})) \\ &= g(f(\boldsymbol{x}_0)) + dg(f(\boldsymbol{x}_0))[df(\boldsymbol{x}_0)(\boldsymbol{x} - \boldsymbol{x}_0) + r_1(\boldsymbol{x})] + r_2(f(\boldsymbol{x})) \\ &= g(f(\boldsymbol{x}_0)) + [dg(f(\boldsymbol{x}_0)) \circ df(\boldsymbol{x}_0)](\boldsymbol{x} - \boldsymbol{x}_0) + r_3(\boldsymbol{x})\,, \end{aligned}$$

where

$$\begin{aligned} r_3(\boldsymbol{x}) &= dg(f(\boldsymbol{x}_0))(r_1(\boldsymbol{x})) + r_2(f(\boldsymbol{x})) \\ &= dg(f(\boldsymbol{x}_0))(r_1(\boldsymbol{x})) + \|f(\boldsymbol{x}) - f(\boldsymbol{x}_0)\| R_2(f(\boldsymbol{x})) \\ &= dg(f(\boldsymbol{x}_0))(r_1(\boldsymbol{x})) + \|df(\boldsymbol{x}_0)(\boldsymbol{x} - \boldsymbol{x}_0) + r_1(\boldsymbol{x})\| R_2(f(\boldsymbol{x}))\,. \end{aligned}$$

Hence,

$$\begin{aligned} \frac{\|r_3(\boldsymbol{x})\|}{\|\boldsymbol{x} - \boldsymbol{x}_0\|} &\leq \left\| dg(f(\boldsymbol{x}_0)) \left(\frac{r_1(\boldsymbol{x})}{\|\boldsymbol{x} - \boldsymbol{x}_0\|} \right) \right\| + \\ &\quad + \left(\left\| df(\boldsymbol{x}_0) \left(\frac{\boldsymbol{x} - \boldsymbol{x}_0}{\|\boldsymbol{x} - \boldsymbol{x}_0\|} \right) \right\| + \frac{\|r_1(\boldsymbol{x})\|}{\|\boldsymbol{x} - \boldsymbol{x}_0\|} \right) \|R_2(f(\boldsymbol{x}))\|\,. \end{aligned}$$

If $x \to x_0$, the first summand tends to 0, since $dg(f(x_0))$ is continuous. Since f is continuous at x_0 and R_2 is continuous at $y_0 = f(x_0)$, with $R_2(y_0) = \mathbf{0}$, we have that $\|R_2(f(x))\|$ tends to 0; on the other hand $df(x_0)$, being continuous, is bounded on the compact set $\overline{B}(\mathbf{0}, 1)$. Therefore,

$$\lim_{x \to x_0} \frac{\|r_3(x)\|}{\|x - x_0\|} = 0 .$$

We conclude that $g \circ f$ is differentiable at x_0, with differential $dg(f(x_0)) \circ df(x_0)$. □

It is well-known that the matrix associated to the composite of two linear functions is the product of the two respective matrices. From the above theorem we then have the following formula for the Jacobian matrices:

$$J(g \circ f)(x_0) = Jg(f(x_0)) \cdot Jf(x_0) ,$$

i.e., the matrix

$$\begin{pmatrix} \frac{\partial (g\circ f)_1}{\partial x_1}(x_0) & \cdots & \frac{\partial (g\circ f)_1}{\partial x_N}(x_0) \\ \vdots & \cdots & \vdots \\ \frac{\partial (g\circ f)_L}{\partial x_1}(x_0) & \cdots & \frac{\partial (g\circ f)_L}{\partial x_N}(x_0) \end{pmatrix}$$

is equal to the product

$$\begin{pmatrix} \frac{\partial g_1}{\partial y_1}(f(x_0)) & \cdots & \frac{\partial g_1}{\partial y_M}(f(x_0)) \\ \vdots & \cdots & \vdots \\ \frac{\partial g_L}{\partial y_1}(f(x_0)) & \cdots & \frac{\partial g_L}{\partial y_M}(f(x_0)) \end{pmatrix} \begin{pmatrix} \frac{\partial f_1}{\partial x_1}(x_0) & \cdots & \frac{\partial f_1}{\partial x_N}(x_0) \\ \vdots & \cdots & \vdots \\ \frac{\partial f_M}{\partial x_1}(x_0) & \cdots & \frac{\partial f_M}{\partial x_N}(x_0) \end{pmatrix} .$$

We can thus derive the formula for the partial derivatives, usually called **chain rule**:

$$\begin{aligned} \frac{\partial (g \circ f)_i}{\partial x_k}(x_0) &= \\ &= \frac{\partial g_i}{\partial y_1}(f(x_0)) \frac{\partial f_1}{\partial x_k}(x_0) + \frac{\partial g_i}{\partial y_2}(f(x_0)) \frac{\partial f_2}{\partial x_k}(x_0) + \cdots + \frac{\partial g_i}{\partial y_M}(f(x_0)) \frac{\partial f_M}{\partial x_k}(x_0) \\ &= \sum_{j=1}^{M} \frac{\partial g_i}{\partial y_j}(f(x_0)) \frac{\partial f_j}{\partial x_k}(x_0) , \end{aligned}$$

where $i = 1, 2, \ldots, L$ and $k = 1, 2, \ldots, N$.

A.5 The Implicit Function Theorem

Let Ω be an open subset of $\mathbb{R}^M \times \mathbb{R}^N$, and $g : \Omega \to \mathbb{R}^N$ be a $\mathcal{C}^1$-function. Hence, g has N components

$$g(x, y) = (g_1(x, y), \ldots, g_N(x, y)) .$$

Here $\boldsymbol{x} = (x_1, \ldots, x_M) \in \mathbb{R}^M$, and $\boldsymbol{y} = (y_1, \ldots, y_N) \in \mathbb{R}^N$. We will use the following notation for the Jacobian matrices:

$$\frac{\partial g}{\partial \boldsymbol{x}}(\boldsymbol{x}, \boldsymbol{y}) = \begin{pmatrix} \frac{\partial g_1}{\partial x_1}(\boldsymbol{x}, \boldsymbol{y}) & \cdots & \frac{\partial g_1}{\partial x_M}(\boldsymbol{x}, \boldsymbol{y}) \\ \vdots & \cdots & \vdots \\ \frac{\partial g_N}{\partial x_1}(\boldsymbol{x}, \boldsymbol{y}) & \cdots & \frac{\partial g_N}{\partial x_M}(\boldsymbol{x}, \boldsymbol{y}) \end{pmatrix}, \quad \frac{\partial g}{\partial \boldsymbol{y}}(\boldsymbol{x}, \boldsymbol{y}) = \begin{pmatrix} \frac{\partial g_1}{\partial y_1}(\boldsymbol{x}, \boldsymbol{y}) & \cdots & \frac{\partial g_1}{\partial y_N}(\boldsymbol{x}, \boldsymbol{y}) \\ \vdots & \cdots & \vdots \\ \frac{\partial g_N}{\partial y_1}(\boldsymbol{x}, \boldsymbol{y}) & \cdots & \frac{\partial g_N}{\partial y_N}(\boldsymbol{x}, \boldsymbol{y}) \end{pmatrix}.$$

Theorem A.8

Let $\Omega \subseteq \mathbb{R}^M \times \mathbb{R}^N$ be open, $g : \Omega \to \mathbb{R}^N$ a $\mathcal{C}^1$-function, and $(\boldsymbol{x}_0, \boldsymbol{y}_0)$ a point in Ω for which

$$g(\boldsymbol{x}_0, \boldsymbol{y}_0) = \boldsymbol{0}, \qquad \text{and} \qquad \det \frac{\partial g}{\partial \boldsymbol{y}}(\boldsymbol{x}_0, \boldsymbol{y}_0) \neq 0.$$

Then, there exist an open neighborhood U of $\boldsymbol{x}_0$, an open neighborhood V of $\boldsymbol{y}_0$, and a $\mathcal{C}^1$-function $\eta : U \to V$ such that $U \times V \subseteq \Omega$ and, taking $\boldsymbol{x} \in U$ and $\boldsymbol{y} \in V$, one has that

$$g(\boldsymbol{x}, \boldsymbol{y}) = \boldsymbol{0} \quad \Longleftrightarrow \quad \boldsymbol{y} = \eta(\boldsymbol{x}).$$

Moreover, the function η is of class $\mathcal{C}^1$, and the following formula holds true:

$$J\eta(\boldsymbol{x}) = -\left(\frac{\partial g}{\partial \boldsymbol{y}}(\boldsymbol{x}, \eta(\boldsymbol{x}))\right)^{-1} \frac{\partial g}{\partial \boldsymbol{x}}(\boldsymbol{x}, \eta(\boldsymbol{x})).$$

Proof

It will be carried out by induction. We first prove the case $N = 1$.[2]

Assume for instance that $\frac{\partial g}{\partial y}(\boldsymbol{x}_0, y_0) > 0$. By the continuity of $\frac{\partial g}{\partial y}$, there is a $\delta > 0$ such that, if $\|\boldsymbol{x} - \boldsymbol{x}_0\| \leq \delta$ and $|y - y_0| \leq \delta$, then $\frac{\partial g}{\partial y}(\boldsymbol{x}, y) > 0$. Hence, for every $\boldsymbol{x} \in \overline{B}(\boldsymbol{x}_0, \delta)$, the function $g(\boldsymbol{x}, \cdot)$ is strictly increasing on $[y_0 - \delta, y_0 + \delta]$. Being $g(\boldsymbol{x}_0, y_0) = 0$, we have that

$$g(\boldsymbol{x}_0, y_0 - \delta) < 0 < g(\boldsymbol{x}_0, y_0 + \delta).$$

[2]The proof reported here is most probably due to Giuseppe Peano, and can be found in the Italian book "Calcolo differenziale e principii di calcolo integrale", published in 1884. This important work was written by Peano himself, who at that time was only 25 years old, but the official author is Angelo Genocchi, the professor who was in care of Peano as an assistant at the University of Torino. Genocchi got indeed very angry when he was told that the book had been published, and publicly declared that he was not aware of what had been written in the volume, recalling however the fact that Peano had followed his lessons in infinitesimal analysis.

By continuity again, there is a $\delta' > 0$ such that, if $\boldsymbol{x} \in \overline{B}(\boldsymbol{x}_0, \delta')$, then

$$g(\boldsymbol{x}, y_0 - \delta) < 0 < g(\boldsymbol{x}, y_0 + \delta).$$

We define $U = B(\boldsymbol{x}_0, \delta')$, and $V =]y_0 - \delta, y_0 + \delta[$. Hence, for every $\boldsymbol{x} \in U$, since $g(\boldsymbol{x}, \cdot)$ is strictly increasing, there is exactly one $y \in]y_0 - \delta, y_0 + \delta[$ for which $g(\boldsymbol{x}, y) = 0$; we call $\eta(\boldsymbol{x})$ such a y. We have thus defined a function $\eta : U \to V$ such that, taking $\boldsymbol{x} \in U$ and $y \in V$,

$$g(\boldsymbol{x}, y) = 0 \quad \Longleftrightarrow \quad y = \eta(\boldsymbol{x}).$$

In order to verify the continuity of η, let us fix a $\bar{\boldsymbol{x}} \in U$ and prove that η is continuous at $\bar{\boldsymbol{x}}$. Taken $\boldsymbol{x} \in U$ and considered the function $\gamma : [0, 1] \to U \times V$, defined as

$$\gamma(t) = (\bar{\boldsymbol{x}} + t(\boldsymbol{x} - \bar{\boldsymbol{x}}), \eta(\bar{\boldsymbol{x}}) + t(\eta(\boldsymbol{x}) - \eta(\bar{\boldsymbol{x}}))),$$

the Lagrange Mean Value Theorem applied to $g \circ \gamma$ tells us that there is a $\xi \in]0, 1[$ for which

$$g(\boldsymbol{x}, \eta(\boldsymbol{x})) - g(\bar{\boldsymbol{x}}, \eta(\bar{\boldsymbol{x}})) = \frac{\partial g}{\partial \boldsymbol{x}}(\gamma(\xi))(\boldsymbol{x} - \bar{\boldsymbol{x}}) + \frac{\partial g}{\partial y}(\gamma(\xi))(\eta(\boldsymbol{x}) - \eta(\bar{\boldsymbol{x}})).$$

Being $g(\boldsymbol{x}, \eta(\boldsymbol{x})) = g(\bar{\boldsymbol{x}}, \eta(\bar{\boldsymbol{x}})) = 0$, we have that

$$|\eta(\boldsymbol{x}) - \eta(\bar{\boldsymbol{x}})| = \frac{1}{|\frac{\partial g}{\partial y}(\gamma(\xi))|} \left\| \frac{\partial g}{\partial \boldsymbol{x}}(\gamma(\xi))(\boldsymbol{x} - \bar{\boldsymbol{x}}) \right\|.$$

Since the partial derivatives of g are continuous and $\frac{\partial g}{\partial y}$ in not zero on the compact set $\overline{U} \times \overline{V}$, we have that there is a constant $c > 0$ for which

$$\frac{1}{|\frac{\partial g}{\partial y}(\gamma(\xi))|} \left\| \frac{\partial g}{\partial \boldsymbol{x}}(\gamma(\xi))(\boldsymbol{x} - \bar{\boldsymbol{x}}) \right\| \le c\|\boldsymbol{x} - \bar{\boldsymbol{x}}\|.$$

As a consequence, η is continuous at $\bar{\boldsymbol{x}}$.

We now prove the differentiability. Taken $\bar{\boldsymbol{x}} = (\bar{x}_1, \bar{x}_2, \dots, \bar{x}_M)$, let $\boldsymbol{x} = (\bar{x}_1 + h, \bar{x}_2, \dots, \bar{x}_M)$; proceeding as above, for h small enough we have

$$\frac{\eta(\bar{x}_1 + h, \bar{x}_2, \dots, \bar{x}_M) - \eta(\bar{x}_1, \bar{x}_2, \dots, \bar{x}_M)}{h} = -\frac{\frac{\partial g}{\partial x_1}(\gamma(\xi_h))}{\frac{\partial g}{\partial y}(\gamma(\xi_h))},$$

with $\gamma(\xi_h)$ belonging to the segment joining $(\bar{\boldsymbol{x}}, \eta(\bar{\boldsymbol{x}}))$ to $(\boldsymbol{x}, \eta(\boldsymbol{x}))$. If h tends to 0, we have that $\gamma(\xi_h)$ tends to $(\bar{\boldsymbol{x}}, \eta(\bar{\boldsymbol{x}}))$, and hence

$$\frac{\partial \eta}{\partial x_1}(\bar{\boldsymbol{x}}) = \lim_{h \to 0} \frac{\eta(\bar{x}_1 + h, \bar{x}_2, \dots, \bar{x}_M) - \eta(\bar{x}_1, \bar{x}_2, \dots, \bar{x}_M)}{h} = -\frac{\frac{\partial g}{\partial x_1}(\bar{\boldsymbol{x}}, \eta(\bar{\boldsymbol{x}}))}{\frac{\partial g}{\partial y}(\bar{\boldsymbol{x}}, \eta(\bar{\boldsymbol{x}}))}.$$

The partial derivatives with respect to $x_2, \dots, x_M$ are computed similarly, thus yielding that η is of class $\mathcal{C}^1$, and

$$J\eta(\boldsymbol{x}) = -\frac{1}{\frac{\partial g}{\partial y}(\boldsymbol{x}, \eta(\boldsymbol{x}))} \frac{\partial g}{\partial \boldsymbol{x}}(\boldsymbol{x}, \eta(\boldsymbol{x})) .$$

We now assume that the statement holds till $N-1$, for some $N \geq 2$ (and any $M \geq 1$), and prove that it then also holds for N. We will use the notation

$$\boldsymbol{y}_1 = (y_1, \dots, y_{N-1}) ,$$

and we will write $\boldsymbol{y} = (\boldsymbol{y}_1, y_N)$. Since

$$\det \begin{pmatrix} \frac{\partial g_1}{\partial y_1}(\boldsymbol{x}_0, \boldsymbol{y}_0) & \cdots & \frac{\partial g_1}{\partial y_N}(\boldsymbol{x}_0, \boldsymbol{y}_0) \\ \vdots & \cdots & \vdots \\ \frac{\partial g_N}{\partial y_1}(\boldsymbol{x}_0, \boldsymbol{y}_0) & \cdots & \frac{\partial g_N}{\partial y_N}(\boldsymbol{x}_0, \boldsymbol{y}_0) \end{pmatrix} \neq 0 ,$$

at least one of the elements in the last column is different from zero. We can assume without loss of generality, possibly changing the rows, that $\frac{\partial g_N}{\partial y_N}(\boldsymbol{x}_0, \boldsymbol{y}_0) \neq 0$. Writing $\boldsymbol{y}_0 = (\boldsymbol{y}_1^0, y_N^0)$, with $\boldsymbol{y}_1^0 = (y_1^0, \dots, y_{N-1}^0)$, we then have

$$g_N(\boldsymbol{x}_0, \boldsymbol{y}_1^0, y_N^0) = 0 , \qquad \text{and} \qquad \frac{\partial g_N}{\partial y_N}(\boldsymbol{x}_0, \boldsymbol{y}_1^0, y_N^0) \neq 0 .$$

Then, by the already proved one-dimensional case, there are an open neighborhood U_1 of $(\boldsymbol{x}_0, \boldsymbol{y}_1^0)$, an open neighborhood V_N of y_N^0, and a $\mathcal{C}^1$-function $\eta_1 : U_1 \to V_N$ such that $U_1 \times V_N \subseteq \Omega$, with the following properties: if $(\boldsymbol{x}, \boldsymbol{y}_1) \in U_1$ and $y_N \in V_N$,

$$g_N(\boldsymbol{x}, \boldsymbol{y}_1, y_N) = 0 \iff y_N = \eta_1(\boldsymbol{x}, \boldsymbol{y}_1) ,$$

and

$$J\eta_1(\boldsymbol{x}, \boldsymbol{y}_1) = -\frac{1}{\frac{\partial g_N}{\partial y_N}(\boldsymbol{x}, \boldsymbol{y}_1, \eta_1(\boldsymbol{x}, \boldsymbol{y}_1)))} \frac{\partial g_N}{\partial(\boldsymbol{x}, \boldsymbol{y}_1)}(\boldsymbol{x}, \boldsymbol{y}_1, \eta_1(\boldsymbol{x}, \boldsymbol{y}_1)) .$$

We may assume U_1 to be of the type $\widetilde{U} \times \widetilde{V}_1$, with $\widetilde{U}$ being an open neighborhood of $\boldsymbol{x}_0$ and $\widetilde{V}_1$ an open neighborhood of $\boldsymbol{y}_1^0$.

Let us define the function $\phi : \widetilde{U} \times \widetilde{V}_1 \to \mathbb{R}^{N-1}$ by setting

$$\phi(\boldsymbol{x}, \boldsymbol{y}_1) = (g_1(\boldsymbol{x}, \boldsymbol{y}_1, \eta_1(\boldsymbol{x}, \boldsymbol{y}_1)), \dots, g_{N-1}(\boldsymbol{x}, \boldsymbol{y}_1, \eta_1(\boldsymbol{x}, \boldsymbol{y}_1))) .$$

For briefness, we will write

$$g_{(1,\dots,N-1)}(\boldsymbol{x}, \boldsymbol{y}) = (g_1(\boldsymbol{x}, \boldsymbol{y}), \dots, g_{N-1}(\boldsymbol{x}, \boldsymbol{y})) ,$$

so that

$$\phi(\boldsymbol{x}, \boldsymbol{y}_1) = g_{(1,\dots,N-1)}(\boldsymbol{x}, \boldsymbol{y}_1, \eta_1(\boldsymbol{x}, \boldsymbol{y}_1)).$$

Notice that, being $\eta_1(\boldsymbol{x}_0, \boldsymbol{y}_1^0) = y_N^0$, we have that

$$\phi(\boldsymbol{x}_0, \boldsymbol{y}_1^0) = g_{(1,\dots,N-1)}(\boldsymbol{x}_0, \boldsymbol{y}_0) = \boldsymbol{0},$$

and

$$\frac{\partial \phi}{\partial \boldsymbol{y}_1}(\boldsymbol{x}_0, \boldsymbol{y}_1^0) = \frac{\partial g_{(1,\dots,N-1)}}{\partial \boldsymbol{y}_1}(\boldsymbol{x}_0, \boldsymbol{y}_0) + \frac{\partial g_{(1,\dots,N-1)}}{\partial y_N}(\boldsymbol{x}_0, \boldsymbol{y}_0)\frac{\partial \eta_1}{\partial \boldsymbol{y}_1}(\boldsymbol{x}_0, \boldsymbol{y}_1^0). \qquad (*)$$

Moreover, since $g_N(\boldsymbol{x}, \boldsymbol{y}_1, \eta_1(\boldsymbol{x}, \boldsymbol{y}_1)) = 0$, for every $(\boldsymbol{x}, \boldsymbol{y}_1) \in U_1$, differentiating with respect to $\boldsymbol{y}_1$ we see that

$$0 = \frac{\partial g_N}{\partial \boldsymbol{y}_1}(\boldsymbol{x}_0, \boldsymbol{y}_0) + \frac{\partial g_N}{\partial y_N}(\boldsymbol{x}_0, \boldsymbol{y}_0)\frac{\partial \eta_1}{\partial \boldsymbol{y}_1}(\boldsymbol{x}_0, \boldsymbol{y}_1^0). \qquad (**)$$

Let us write the identity

$$\det \frac{\partial \phi}{\partial \boldsymbol{y}_1}(\boldsymbol{x}_0, \boldsymbol{y}_1^0) = \frac{1}{\frac{\partial g_N}{\partial y_N}(\boldsymbol{x}_0, \boldsymbol{y}_0)} \det \left(\begin{array}{c|c} \frac{\partial \phi}{\partial \boldsymbol{y}_1}(\boldsymbol{x}_0, \boldsymbol{y}_1^0) & \frac{\partial g_{(1,\dots,N-1)}}{\partial y_N}(\boldsymbol{x}_0, \boldsymbol{y}_0) \\ \hline 0 & \frac{\partial g_N}{\partial y_N}(\boldsymbol{x}_0, \boldsymbol{y}_0) \end{array} \right),$$

Substituting the two equalities (*), (**), we have that

$$\det \left(\begin{array}{c|c} \frac{\partial \phi}{\partial \boldsymbol{y}_1}(\boldsymbol{x}_0, \boldsymbol{y}_1^0) & \frac{\partial g_{(1,\dots,N-1)}}{\partial y_N}(\boldsymbol{x}_0, \boldsymbol{y}_0) \\ \hline 0 & \frac{\partial g_N}{\partial y_N}(\boldsymbol{x}_0, \boldsymbol{y}_0) \end{array} \right) =$$

$$= \det \left(\begin{array}{c|c} \frac{\partial g_{(1,\dots,N-1)}}{\partial \boldsymbol{y}_1}(\boldsymbol{x}_0, \boldsymbol{y}_0) + \frac{\partial g_{(1,\dots,N-1)}}{\partial y_N}(\boldsymbol{x}_0, \boldsymbol{y}_0)\frac{\partial \eta_1}{\partial \boldsymbol{y}_1}(\boldsymbol{x}_0, \boldsymbol{y}_1^0) & \frac{\partial g_{(1,\dots,N-1)}}{\partial y_N}(\boldsymbol{x}_0, \boldsymbol{y}_0) \\ \hline \frac{\partial g_N}{\partial \boldsymbol{y}_1}(\boldsymbol{x}_0, \boldsymbol{y}_0) + \frac{\partial g_N}{\partial y_N}(\boldsymbol{x}_0, \boldsymbol{y}_0)\frac{\partial \eta_1}{\partial \boldsymbol{y}_1}(\boldsymbol{x}_0, \boldsymbol{y}_1^0) & \frac{\partial g_N}{\partial y_N}(\boldsymbol{x}_0, \boldsymbol{y}_0) \end{array} \right)$$

$$= \det \left(\frac{\partial g}{\partial \boldsymbol{y}_1}(\boldsymbol{x}_0, \boldsymbol{y}_0) + \frac{\partial g}{\partial y_N}(\boldsymbol{x}_0, \boldsymbol{y}_0)\frac{\partial \eta_1}{\partial \boldsymbol{y}_1}(\boldsymbol{x}_0, \boldsymbol{y}_1^0) \,\middle|\, \frac{\partial g}{\partial y_N}(\boldsymbol{x}_0, \boldsymbol{y}_0) \right)$$

$$= \det \left[\frac{\partial g}{\partial \boldsymbol{y}}(\boldsymbol{x}_0, \boldsymbol{y}_0) + \left(\frac{\partial g}{\partial y_N}(\boldsymbol{x}_0, \boldsymbol{y}_0)\frac{\partial \eta_1}{\partial \boldsymbol{y}_1}(\boldsymbol{x}_0, \boldsymbol{y}_1^0) \,\middle|\, 0 \right) \right].$$

We now recall that, adding a scalar multiple of one column to another column of a matrix does not change the value of its determinant. Hence, being

$$\left(\frac{\partial g}{\partial y_N}(\boldsymbol{x}_0, \boldsymbol{y}_0)\frac{\partial \eta_1}{\partial \boldsymbol{y}_1}(\boldsymbol{x}_0, \boldsymbol{y}_1^0)\,\Big|\, 0\right) =$$

$$= \begin{pmatrix} \frac{\partial g_1}{\partial y_N}(\boldsymbol{x}_0, \boldsymbol{y}_0)\frac{\partial \eta_1}{\partial y_1}(\boldsymbol{x}_0, \boldsymbol{y}_1^0) & \cdots & \frac{\partial g_1}{\partial y_N}(\boldsymbol{x}_0, \boldsymbol{y}_0)\frac{\partial \eta_1}{\partial y_{N-1}}(\boldsymbol{x}_0, \boldsymbol{y}_1^0) & 0 \\ \vdots & \cdots & \vdots & \vdots \\ \frac{\partial g_N}{\partial y_N}(\boldsymbol{x}_0, \boldsymbol{y}_0)\frac{\partial \eta_1}{\partial y_1}(\boldsymbol{x}_0, \boldsymbol{y}_1^0) & \cdots & \frac{\partial g_N}{\partial y_N}(\boldsymbol{x}_0, \boldsymbol{y}_0)\frac{\partial \eta_1}{\partial y_{N-1}}(\boldsymbol{x}_0, \boldsymbol{y}_1^0) & 0 \end{pmatrix},$$

each of its columns is a scalar multiple of $\frac{\partial g}{\partial y_N}(\boldsymbol{x}_0, \boldsymbol{y}_0)$, hence

$$\det\left[\frac{\partial g}{\partial \boldsymbol{y}}(\boldsymbol{x}_0, \boldsymbol{y}_0) + \left(\frac{\partial g}{\partial y_N}(\boldsymbol{x}_0, \boldsymbol{y}_0)\frac{\partial \eta_1}{\partial \boldsymbol{y}_1}(\boldsymbol{x}_0, \boldsymbol{y}_1^0)\,\Big|\, 0\right)\right] = \det \frac{\partial g}{\partial \boldsymbol{y}}(\boldsymbol{x}_0, \boldsymbol{y}_0)\,.$$

So, finally we have

$$\phi(\boldsymbol{x}_0, \boldsymbol{y}_1^0) = \boldsymbol{0}\,, \qquad \text{and} \qquad \det \frac{\partial \phi}{\partial \boldsymbol{y}_1}(\boldsymbol{x}_0, \boldsymbol{y}_1^0) \neq 0\,.$$

By the inductive assumption, there are an open neighborhood U of $\boldsymbol{x}_0$, an open neighborhood V_1 of $\boldsymbol{y}_1^0$ and a $\mathcal{C}^1$-function $\eta_2 : U \to V_1$ such that $U \times V_1 \subseteq \widetilde{U} \times \widetilde{V}_1$, and the following holds: for every $\boldsymbol{x} \in U$ and $\boldsymbol{y}_1 \in V_1$,

$$\phi(\boldsymbol{x}, \boldsymbol{y}_1) = \boldsymbol{0} \quad \Longleftrightarrow \quad \boldsymbol{y}_1 = \eta_2(\boldsymbol{x})\,.$$

In conclusion, for $\boldsymbol{x} \in U$ and $\boldsymbol{y} = (\boldsymbol{y}_1, y_N) \in V_1 \times V_2$, we have that

$$\begin{aligned} g(\boldsymbol{x}, \boldsymbol{y}) = \boldsymbol{0} \quad &\Longleftrightarrow \quad \begin{cases} g_{(1,\dots,N-1)}(\boldsymbol{x}, \boldsymbol{y}_1, y_N) = \boldsymbol{0} \\ g_N(\boldsymbol{x}, \boldsymbol{y}_1, y_N) = 0 \end{cases} \\ &\Longleftrightarrow \quad \begin{cases} g_{(1,\dots,N-1)}(\boldsymbol{x}, \boldsymbol{y}_1, y_N) = \boldsymbol{0} \\ y_N = \eta_1(\boldsymbol{x}, \boldsymbol{y}_1) \end{cases} \\ &\Longleftrightarrow \quad \begin{cases} \phi(\boldsymbol{x}, \boldsymbol{y}_1) = \boldsymbol{0} \\ y_N = \eta_1(\boldsymbol{x}, \boldsymbol{y}_1) \end{cases} \\ &\Longleftrightarrow \quad \begin{cases} \boldsymbol{y}_1 = \eta_2(\boldsymbol{x}) \\ y_N = \eta_1(\boldsymbol{x}, \boldsymbol{y}_1) \end{cases} \\ &\Longleftrightarrow \quad \boldsymbol{y} = (\eta_2(\boldsymbol{x}), \eta_1(\boldsymbol{x}, \eta_2(\boldsymbol{x})))\,. \end{aligned}$$

Setting $V = V_1 \times V_2$, we may then define the function $\eta : U \to V$ as

$$\eta(\boldsymbol{x}) = (\eta_2(\boldsymbol{x}), \eta_1(\boldsymbol{x}, \eta_2(\boldsymbol{x}))) .$$

This function is of class $\mathcal{C}^1$, since both η_1 and η_2 are such. Since $g(\boldsymbol{x}, \eta(\boldsymbol{x})) = 0$ for every $\boldsymbol{x} \in U$, one easily deduces that

$$\frac{\partial g}{\partial \boldsymbol{x}}(\boldsymbol{x}, \eta(\boldsymbol{x})) + \frac{\partial g}{\partial \boldsymbol{y}}(\boldsymbol{x}, \eta(\boldsymbol{x})) J\eta(\boldsymbol{x}) = 0 ,$$

whence the formula for $J\eta(\boldsymbol{x})$. □

Clearly, the following analogous statement holds true, where the roles of $\boldsymbol{x}$ and $\boldsymbol{y}$ are interchanged.

Theorem A.9
Let $\Omega \subseteq \mathbb{R}^M \times \mathbb{R}^N$ be open, $g : \Omega \to \mathbb{R}^N$ a $\mathcal{C}^1$-function, and $(\boldsymbol{x}_0, \boldsymbol{y}_0)$ a point in Ω for which

$$g(\boldsymbol{x}_0, \boldsymbol{y}_0) = \boldsymbol{0} , \qquad \textit{and} \qquad \det \frac{\partial g}{\partial \boldsymbol{x}}(\boldsymbol{x}_0, \boldsymbol{y}_0) \neq 0 .$$

Then, there exist an open neighborhood U of $\boldsymbol{x}_0$, an open neighborhood V of $\boldsymbol{y}_0$, and a $\mathcal{C}^1$-function $\eta : V \to U$ such that $U \times V \subseteq \Omega$ and, taking $\boldsymbol{x} \in U$ and $\boldsymbol{y} \in V$, one has that

$$g(\boldsymbol{x}, \boldsymbol{y}) = \boldsymbol{0} \iff \boldsymbol{x} = \eta(\boldsymbol{y}) .$$

Moreover, the function η is of class $\mathcal{C}^1$, and the following formula holds true:

$$J\eta(\boldsymbol{y}) = -\left(\frac{\partial g}{\partial \boldsymbol{x}}(\boldsymbol{y}, \eta(\boldsymbol{y}))\right)^{-1} \frac{\partial g}{\partial \boldsymbol{y}}(\boldsymbol{y}, \eta(\boldsymbol{y})) .$$

A.6 Local Diffeomorphisms

Let us introduce the notion of **diffeomorphism**.

Definition A.10
Given A and B, two open subsets of $\mathbb{R}^N$, a function $\varphi : A \to B$ is said to be a diffeomorphism if it is of class $\mathcal{C}^1$, it is a bijection, and its inverse $\varphi^{-1} : B \to A$ is also of class $\mathcal{C}^1$.

Let us state the important **Local Diffeomorphism Theorem**.

Theorem A.11
Let A and B be open subsets of $\mathbb{R}^N$, and $\varphi : A \to B$ be a $\mathcal{C}^1$-function. If, for some $\boldsymbol{x}_0 \in A$, one has that $\det J\varphi(\boldsymbol{x}_0) \neq 0$, *then there exist an open neighborhood U of $\boldsymbol{x}_0$ contained in A, and an open neighborhood V of $\varphi(\boldsymbol{x}_0)$ contained in B, such that the restricted function $\varphi_{|U} : U \to V$ is a diffeomorfism.*

Proof
We consider the function $g : A \times B \to \mathbb{R}^N$ defined as

$$g(\boldsymbol{x}, \boldsymbol{y}) = \boldsymbol{y} - \varphi(\boldsymbol{x}).$$

Setting $\boldsymbol{y}_0 = \varphi(\boldsymbol{x}_0)$, we have that

$$g(\boldsymbol{x}_0, \boldsymbol{y}_0) = \boldsymbol{0}, \qquad \text{and} \qquad \det \frac{\partial g}{\partial \boldsymbol{x}}(\boldsymbol{x}_0, \boldsymbol{y}_0) = \det J\varphi(\boldsymbol{x}_0) \neq 0.$$

By the Implicit Function Theorem, there exist an open neighborhood V of $\boldsymbol{y}_0$, an open neighborhood U of $\boldsymbol{x}_0$ and a $\mathcal{C}^1$ function $\eta : V \to U$ such that, taken $\boldsymbol{y} \in V$ and $\boldsymbol{x} \in U$,

$$\varphi(\boldsymbol{x}) = \boldsymbol{y} \quad \Longleftrightarrow \quad g(\boldsymbol{x}, \boldsymbol{y}) = \boldsymbol{0} \quad \Longleftrightarrow \quad \boldsymbol{x} = \eta(\boldsymbol{y}).$$

Hence, $\eta = \varphi_{|U}^{-1}$, and the proof is thus completed. □

Appendix B
Stokes–Cartan and Poincaré Theorems

Alessandro Fonda

A. Fonda, *The Kurzweil-Henstock Integral for Undergraduates*,
Compact Textbooks in Mathematics,
https://doi.org/10.1007/978-3-319-95321-2

In this appendix we will give a complete proof of the Stokes–Cartan and Poincaré theorems, of which only particular cases have been proved in ▶ Chap. 3.

Let U be an open set in $\mathbb{R}^N$, V an open set in[1] $\mathbb{R}^P$ and $\phi : V \to U$ a function of class $\mathcal{C}^1$:

$$\phi(\boldsymbol{y}) = (\phi_1(\boldsymbol{y}), \dots, \phi_N(\boldsymbol{y})),$$

with $\boldsymbol{y} = (y_1, \dots, y_P) \in V$. Given a M-differential form $\omega : U \to \Omega_M(\mathbb{R}^N)$,

$$\omega(\boldsymbol{x}) = \sum_{1 \le i_1 < \dots < i_M \le N} f_{i_1,\dots,i_M}(\boldsymbol{x})\, dx_{i_1} \wedge \dots \wedge dx_{i_M},$$

we can define a M-differential form $T_\phi \omega : V \to \Omega_M(\mathbb{R}^P)$, which we will call the transformation through ϕ of ω, in the following way[2]:

$$T_\phi \omega(\boldsymbol{y}) = \sum_{1 \le i_1 < \dots < i_M \le N} f_{i_1,\dots,i_M}(\phi(\boldsymbol{y}))\, d\phi_{i_1}(\boldsymbol{y}) \wedge \dots \wedge d\phi_{i_M}(\boldsymbol{y}).$$

Notice that

$$\begin{aligned} d\phi_{i_1}(\boldsymbol{y}) \wedge \dots \wedge\ d\phi_{i_M}(\boldsymbol{y}) &= \\ &= \left(\sum_{j=1}^{P} \frac{\partial \phi_{i_1}}{\partial y_j}(\boldsymbol{y}) dy_j \right) \wedge \dots \wedge \left(\sum_{j=1}^{P} \frac{\partial \phi_{i_M}}{\partial y_j}(\boldsymbol{y}) dy_j \right) \\ &= \sum_{j_1,\dots,j_M=1}^{P} \frac{\partial \phi_{i_1}}{\partial y_{j_1}}(\boldsymbol{y}) \cdots \frac{\partial \phi_{i_M}}{\partial y_{j_M}}(\boldsymbol{y})\, dy_{j_1} \wedge \dots \wedge dy_{j_M} \end{aligned}$$

[1]Whenever the sets would not be open, see the footnote in ▶ Sect. 3.6.
[2]This is usually called **pull-back** and denoted by $\phi * \omega$.

(being aware that here the indices $j_1, \ldots, j_M$ are not in an increasing order). It is readily verified that, taken a $c \in \mathbb{R}$, it is

$$T_\phi(c\omega) = c\, T_\phi \omega\,;$$

if $\widetilde{\omega}$ is a $\tilde{M}$-differential form defined on U, then

$$T_\phi(\omega \wedge \widetilde{\omega}) = T_\phi \omega \wedge T_\phi \widetilde{\omega}\,,$$

and if moreover $M = \tilde{M}$, then

$$T_\phi(\omega + \widetilde{\omega}) = T_\phi \omega + T_\phi \widetilde{\omega}\,.$$

Let us prove now the following properties.

Proposition B.1
If $\psi : W \to V$ *and* $\phi : V \to U$, *then*

$$T_\psi(T_\phi \omega) = T_{\phi \circ \psi}\, \omega\,.$$

Proof
By the linearity properties seen above, it will be sufficient to consider the case of a differential form of the type

$$\omega(\boldsymbol{x}) = f_{i_1,\ldots,i_M}(\boldsymbol{x})\, dx_{i_1} \wedge \cdots \wedge dx_{i_M}\,.$$

Then,

$$T_\psi(T_\phi \omega) = \left[(f_{i_1,\ldots,i_M} \circ \phi) \sum_{j_1,\ldots,j_M=1}^{P} \frac{\partial \phi_{i_1}}{\partial y_{j_1}} \cdots \frac{\partial \phi_{i_M}}{\partial y_{j_M}} \right] \circ \psi\, d\psi_{j_1} \wedge \cdots \wedge d\psi_{j_M}\,.$$

On the other hand,

$$T_{\phi \circ \psi}\omega = (f_{i_1,\ldots,i_M} \circ \phi \circ \psi)\, d(\phi \circ \psi)_{i_1} \wedge \cdots \wedge d(\phi \circ \psi)_{i_M}\,,$$

and being

$$d(\phi \circ \psi)_{i_k} = d(\phi_{i_k} \circ \psi) = \sum_{j=1}^{P} \left(\frac{\partial \phi_{i_k}}{\partial y_j} \circ \psi \right) d\psi_j\,,$$

equality then holds. □

Proposition B.2
Assume that ϕ be of class $\mathcal{C}^2$. If ω is of class $\mathcal{C}^1$, then also $T_\phi\omega$ is such, and

$$d(T_\phi\omega) = T_\phi(d\omega)\,.$$

Proof
Here, too, it is sufficient to consider the case $\omega = f_{i_1,\dots,i_M}\, dx_{i_1} \wedge \cdots \wedge dx_{i_M}$. We have

$$\begin{aligned} d(T_\phi\omega) &= d(f_{i_1,\dots,i_M} \circ \phi) \wedge d\phi_{i_1} \wedge \cdots \wedge d\phi_{i_M} + (f_{i_1,\dots,i_M} \circ \phi)\, d(d\phi_{i_1} \wedge \cdots \wedge d\phi_{i_M}) \\ &= d(f_{i_1,\dots,i_M} \circ \phi) \wedge d\phi_{i_1} \wedge \cdots \wedge d\phi_{i_M} \\ &= \left[\sum_{m=1}^{N} \left(\frac{\partial f_{i_1,\dots,i_M}}{\partial x_m} \circ \phi \right) d\phi_m \right] \wedge d\phi_{i_1} \wedge \cdots \wedge d\phi_{i_M}\,. \end{aligned}$$

On the other hand,

$$d\omega(\boldsymbol{x}) = \sum_{m=1}^{N} \frac{\partial f_{i_1,\dots,i_M}}{\partial x_m}(\boldsymbol{x})\, dx_m \wedge dx_{i_1} \wedge \cdots \wedge dx_{i_M}\,,$$

hence

$$T_\phi(d\omega) = \sum_{m=1}^{N} \left(\frac{\partial f_{i_1,\dots,i_M}}{\partial x_m} \circ \phi \right) d\phi_m \wedge d\phi_{i_1} \wedge \cdots \wedge d\phi_{i_M}\,,$$

and the formula is thus proved. □

Proposition B.3
If $\sigma : I \to \mathbb{R}^N$ is a M-surface whose support is contained in U, then

$$\int_\sigma \omega = \int_I T_\sigma \omega\,.$$

Proof

As above, we just consider the case $\omega = f_{i_1,\dots,i_M} dx_{i_1} \wedge \dots \wedge dx_{i_M}$. We have

$$\int_I T_\sigma \omega = \int_I f_{i_1,\dots,i_M}(\sigma(\boldsymbol{u})) \sum_{j_1,\dots,j_M=1}^{M} \frac{\partial \sigma_{i_1}}{\partial u_{j_1}}(\boldsymbol{u}) \dots \frac{\partial \sigma_{i_M}}{\partial u_{j_M}}(\boldsymbol{u})\, du_{j_1} \wedge \dots \wedge du_{j_M}$$

$$= \int_I f_{i_1,\dots,i_M}(\sigma(\boldsymbol{u})) \det \sigma'_{(i_1,\dots,i_M)}(\boldsymbol{u})\, d\boldsymbol{u}$$

$$= \int_\sigma \omega .$$

This completes the proof. □

We are now ready to give the proof of the **Stokes–Cartan theorem**, whose statement, we recall, is the following.

Theorem B.4

Let $0 \le M \le N-1$. If $\omega : U \to \Omega_M(\mathbb{R}^N)$ is a M-differential form of class $\mathcal{C}^1$ and $\sigma : I \to \mathbb{R}^N$ is a $(M+1)$-surface whose support is contained in U, then

$$\int_\sigma d\omega = \int_{\partial\sigma} \omega .$$

Proof

The case $M = 0$ follows from the Fundamental Theorem applied to the function $\omega \circ \sigma : [a,b] \to \mathbb{R}$. Assume now $1 \le M \le N-1$. Being

$$\int_{\sigma\circ\alpha_k^+} \omega = \int_{I_k} T_{\sigma\circ\alpha_k^+}\omega = \int_{I_k} T_{\alpha_k^+}(T_\sigma \omega) = \int_{\alpha_k^+} T_\sigma \omega ,$$

and analogously for β_k^+ we have

$$\int_{\partial\sigma} \omega = \sum_{k=1}^{M+1} (-1)^k \int_{\sigma\circ\alpha_k^+} \omega + \sum_{k=1}^{M+1} (-1)^{k-1} \int_{\sigma\circ\beta_k^+} \omega$$

$$= \sum_{k=1}^{M+1} (-1)^k \int_{\alpha_k^+} T_\sigma \omega + \sum_{k=1}^{M+1} (-1)^{k-1} \int_{\beta_k^+} T_\sigma \omega$$

$$= \int_{\partial I} T_\sigma \omega .$$

If σ is of class C^2, then $T_\sigma \omega$ is of class C^1 and, applying the Gauss formula to $T_\sigma \omega$, we have

$$\int_{\partial I} T_\sigma \omega = \int_I d(T_\sigma \omega).$$

But

$$\int_I d(T_\sigma \omega) = \int_I T_\sigma (d\omega) = \int_\sigma d\omega.$$

Hence, we have see that

$$\int_\sigma d\omega = \int_I d(T_\sigma \omega) = \int_{\partial I} T_\sigma \omega = \int_{\partial \sigma} \omega,$$

and the theorem is proved in this case.

The assumption that $\sigma : I \to \mathbb{R}^N$ be of class C^2 can be eliminated by an approximation procedure: it is possible to construct a sequence $(\sigma_n)_n$ of M-surfaces of class C^2 which converge to σ together with all first order partial derivatives. The Stokes–Cartan formula then holds for such surfaces, and the Dominated Convergence Theorem permits to pass to the limit and conclude. □

Consider now the set $[0, 1] \times U$, whose elements will be denoted by $(t, \boldsymbol{x}) = (t, x_1, \ldots, x_N)$. Let us define the linear operator K which transforms a generic M-differential form $\alpha : [0, 1] \times U \to \Omega_M(R^{N+1})$ in a $(M-1)$-differential form $K(\alpha) : U \to \Omega_{M-1}(R^N)$ in the following way:

a) if $\alpha(t, \boldsymbol{x}) = f(t, \boldsymbol{x})\, dt \wedge dx_{i_1} \wedge \cdots \wedge dx_{i_{M-1}}$ (notice that here the term dt appears), then

$$K(\alpha)(\boldsymbol{x}) = \left(\int_0^1 f(t, \boldsymbol{x})\, dt \right) dx_{i_1} \wedge \cdots \wedge dx_{i_{M-1}};$$

b) if $\alpha(t, \boldsymbol{x}) = f(t, \boldsymbol{x})\, dx_{i_1} \wedge \cdots \wedge dx_{i_M}$ (here the term dt does not appear), then $K(\alpha) = 0$;

c) in all the other cases, K is defined by linearity (for each component of a generic M-differential form α, the term dt may appear or not, and the previous two definitions apply).

We moreover define the functions $\psi, \xi : U \to [0, 1] \times U$ as follows:

$$\psi(x_1, \ldots, x_N) = (0, x_1, \ldots, x_N), \qquad \xi(x_1, \ldots, x_N) = (1, x_1, \ldots, x_N).$$

Lemma B.5
If $\alpha : [0, 1] \times U \to \Omega_M(R^{N+1})$ *is a* M*-differential form of class* C^1*, then*

$$d(K(\alpha)) + K(d\alpha) = T_\xi\alpha - T_\psi\alpha .$$

Proof
Because of the linearity, it will be sufficient to consider the two cases when the differential form α is of one of the two kinds considered in a) and b).

a) If $\alpha(t, \boldsymbol{x}) = f(t, \boldsymbol{x})\, dt \wedge dx_{i_1} \wedge \cdots \wedge dx_{i_{M-1}}$, by the Leibniz rule we have

$$d(K(\alpha))(\boldsymbol{x}) = \sum_{m=1}^{N} \left(\int_0^1 \frac{\partial f}{\partial x_m}(t, \boldsymbol{x})\, dt \right) dx_m \wedge dx_{i_1} \wedge \cdots \wedge dx_{i_{M-1}} ;$$

on the other hand,

$$\begin{aligned} d\alpha(t, \boldsymbol{x}) &= \frac{\partial f}{\partial t}(t, \boldsymbol{x})\, dt \wedge dt \wedge dx_{i_1} \wedge \cdots \wedge dx_{i_{M-1}} + \\ &\quad + \sum_{m=1}^{N} \frac{\partial f}{\partial x_m}(t, \boldsymbol{x})\, dx_m \wedge dt \wedge dx_{i_1} \wedge \cdots \wedge dx_{i_{M-1}} \\ &= -\sum_{m=1}^{N} \frac{\partial f}{\partial x_m}(t, \boldsymbol{x})\, dt \wedge dx_m \wedge dx_{i_1} \wedge \cdots \wedge dx_{i_{M-1}} , \end{aligned}$$

and hence

$$\begin{aligned} K(d\alpha)(\boldsymbol{x}) &= -\sum_{m=1}^{N} \left(\int_0^1 \frac{\partial f}{\partial x_m}(t, \boldsymbol{x})\, dt \right) dx_m \wedge dx_{i_1} \wedge \cdots \wedge dx_{i_{M-1}} \\ &= -d(K(\alpha))(\boldsymbol{x}) . \end{aligned}$$

Moreover, since the first component of ψ and of ξ is constant, it is $T_\psi\alpha = T_\xi\alpha = 0$. Hence, the identity is proved in this case.

b) If $\alpha(t, \boldsymbol{x}) = f(t, \boldsymbol{x})\, dx_{i_1} \wedge \cdots \wedge dx_{i_M}$, it is $K(\alpha) = 0$ and hence $d(K(\alpha)) = 0$; on the other hand,

$$\begin{aligned} d\alpha(t, \boldsymbol{x}) &= \frac{\partial f}{\partial t}(t, \boldsymbol{x})\, dt \wedge dx_{i_1} \wedge \cdots \wedge dx_{i_M} + \\ &\quad + \sum_{m=1}^{N} \frac{\partial f}{\partial x_m}(t, \boldsymbol{x})\, dx_m \wedge dx_{i_1} \wedge \cdots \wedge dx_{i_M} , \end{aligned}$$

and hence

$$K(d\alpha)(\boldsymbol{x}) = \left(\int_0^1 \frac{\partial f}{\partial t}(t,\boldsymbol{x})\,dt\right) dx_{i_1} \wedge \cdots \wedge dx_{i_M}$$
$$= (f(1,\boldsymbol{x}) - f(0,\boldsymbol{x}))\, dx_{i_1} \wedge \cdots \wedge dx_{i_M}\,.$$

Moreover,

$$T_\xi\alpha(\boldsymbol{x}) = f(1,\boldsymbol{x})\, d\xi_{i_1}(\boldsymbol{x}) \wedge \cdots \wedge d\xi_{i_M}(\boldsymbol{x})$$
$$= f(1,\boldsymbol{x})\, dx_{i_1} \wedge \cdots \wedge dx_{i_M}\,,$$
$$T_\psi\alpha(\boldsymbol{x}) = f(0,\boldsymbol{x})\, d\psi_{i_1}(\boldsymbol{x}) \wedge \cdots \wedge d\psi_{i_M}(\boldsymbol{x})$$
$$= f(0,\boldsymbol{x})\, dx_{i_1} \wedge \cdots \wedge dx_{i_M}\,.$$

The formula is thus proved in this case, as well.

□

We can now give the proof of the **Poincaré theorem**, whose statement is recalled below.

Theorem B.6
Let U be an open subset of $\mathbb{R}^N$, star-shaped with respect to a point $\bar{\boldsymbol{x}}$. For $1 \le M \le N$, a M-differential form $\omega : U \to \Omega_M(\mathbb{R}^N)$ of class $\mathcal{C}^1$ is exact if and only if it is closed. In that case, if ω is of the type

$$\omega(\boldsymbol{x}) = \sum_{1 \le i_1 < \cdots < i_M \le N} f_{i_1,\ldots,i_M}(\boldsymbol{x})\, dx_{i_1} \wedge \cdots \wedge dx_{i_M}\,,$$

a $(M-1)$-differential form $\widetilde{\omega}$ such that $d\widetilde{\omega} = \omega$ is given by

$$\widetilde{\omega}(\boldsymbol{x}) = \sum_{1 \le i_1 < \cdots < i_M \le N} \sum_{s=1}^{M} (-1)^{s+1}(x_{i_s} - \bar{x}_{i_s})\cdot$$
$$\cdot\left(\int_0^1 t^{M-1} f_{i_1,\ldots,i_M}(\bar{\boldsymbol{x}} + t(\boldsymbol{x} - \bar{\boldsymbol{x}}))\,dt\right) dx_{i_1} \wedge \cdots \wedge \widehat{dx_{i_s}} \wedge \cdots \wedge dx_{i_M}\,.$$

Proof
To simplify the notations, we can assume without loss of generality that $\bar{\boldsymbol{x}} = (0, 0, \ldots, 0)$; let $\phi : [0,1] \times U \to U$ be defined by

$$\phi(t, x_1, \ldots, x_N) = (tx_1, \ldots, tx_N)\,.$$

Moreover, by the linearity, we may assume, for simplicity, that

$$\omega(\boldsymbol{x}) = f_{i_1,\dots,i_M}(\boldsymbol{x})\,dx_{i_1} \wedge \cdots \wedge dx_{i_M}\,.$$

Consider $T_\phi\omega$, the transformation through ϕ of ω. It is the differential form of degree M defined on $[0,1] \times U$ as follows:

$$\begin{aligned} T_\phi\omega(t,\boldsymbol{x}) &= f_{i_1,\dots,i_M}(t\boldsymbol{x})(x_{i_1}dt + t\,dx_{i_1}) \wedge \cdots \wedge (x_{i_M}dt + t\,dx_{i_M}) \\ &= f_{i_1,\dots,i_M}(t\boldsymbol{x})[t^M dx_{i_1} \wedge \cdots \wedge dx_{i_M} + \\ &\qquad + t^{M-1}\sum_{s=1}^{M}(-1)^{s-1}x_{i_s}dt \wedge dx_{i_1} \wedge \cdots \wedge \widehat{dx_{i_s}} \wedge \cdots \wedge dx_{i_M}]\,. \end{aligned}$$

Observe that

$$\begin{aligned} &K(T_\phi\omega)(\boldsymbol{x}) = \\ &= \left(\int_0^1 f_{i_1,\dots,i_M}(t\boldsymbol{x})t^{M-1}\sum_{s=1}^{M}(-1)^{s+1}x_{i_s}\,dt\right)dx_{i_1} \wedge \cdots \wedge \widehat{dx_{i_s}} \wedge \cdots \wedge dx_{i_M}\,, \end{aligned}$$

so that $\widetilde{\omega} = K(T_\phi\omega)$. We want to prove that $d\widetilde{\omega} = \omega$. Being ω closed, we have

$$K(d(T_\phi\omega)) = K(T_\phi(d\omega)) = K(T_\phi(0)) = K(0) = 0\,.$$

By the preceding lemma,

$$\begin{aligned} d\widetilde{\omega} &= d(K(T_\phi\omega)) \\ &= T_\xi(T_\phi\omega) - T_\psi(T_\phi\omega) - K(d(T_\phi\omega)) \\ &= T_\xi(T_\phi\omega) - T_\psi(T_\phi\omega) \\ &= T_{\phi\circ\xi}\,\omega - T_{\phi\circ\psi}\,\omega\,. \end{aligned}$$

Being $\phi \circ \xi$ the identity function and $\phi \circ \psi$ the null function, we have that $T_{\phi\circ\xi}\,\omega = \omega$ and $T_{\phi\circ\psi}\,\omega = 0$, which concludes the proof. □

We have thus concluded the proof of the two main theorems of the theory, and maybe something should be said about the need for such a theory. Of course, its mathematical beauty would alone justify its existence and development. Nevertheless, such a nice theory also finds a lot of applications in the physical world. The reason of this probably lies in the fact that it was *motivated* by the classical theorems in Electromagnetism and Fluidodynamics, hence the abstract construction lies on some very concrete bases.

So we find an example here of how Physics and Mathematics interact to help and motivate each other, leading to wonderful successful theories which may have a lot of practical implications in our lifes.

Appendix C
On Differentiable Manifolds

Alessandro Fonda

A. Fonda, *The Kurzweil-Henstock Integral for Undergraduates*,
Compact Textbooks in Mathematics,
https://doi.org/10.1007/978-3-319-95321-2

We would like to show here how the theory on differential forms developed in ▶ Chap. 3, and in particular the Stokes–Cartan theorem, can be adapted to the context of differentiable manifolds. However, contrary to our habit, we will not give the complete proofs of all the results of this section; the interested reader will find useful to refer, e.g., to [20]. We consider a subset $\mathcal{M}$ of $\mathbb{R}^N$.

Definition C.1

The set $\mathcal{M}$ is a **M-dimensional differentiable manifold**, with $1 \leq M \leq N$ (or briefly a **M-manifold**) if, taken a point $\boldsymbol{x}$ in $\mathcal{M}$, there are an open neighborhood A of $\boldsymbol{x}$, an open neighborhood B of $\boldsymbol{0}$ in $\mathbb{R}^N$ and a diffeomorphism $\varphi : A \to B$ such that $\varphi(\boldsymbol{x}) = \boldsymbol{0}$ and, either

(*a*) $\varphi(A \cap \mathcal{M}) = \{y = (y_1, \dots, y_N) \in B : y_{M+1} = \dots = y_N = 0\}$,

or

(*b*) $\varphi(A \cap \mathcal{M}) = \{y = (y_1, \dots, y_N) \in B : y_{M+1} = \dots = y_N = 0 \text{ and } y_M \geq 0\}$.

It can be seen that (*a*) and (*b*) cannot hold at the same time. The points $\boldsymbol{x}$ for which (*b*) is verified make up the **boundary** of $\mathcal{M}$, which we denote by $\partial\mathcal{M}$. If $\partial\mathcal{M}$ is empty, we are speaking of a M-manifold without boundary; otherwise, $\mathcal{M}$ is sometimes said to be a M-manifold with boundary.

First of all we notice that the boundary of a differentiable manifold is itself a differentiable manifold, with a lower dimension.

Theorem C.2
The set $\partial\mathcal{M}$ is a $(M-1)$-manifold without boundary:

$$\partial(\partial\mathcal{M}) = \emptyset\,.$$

Proof
Taken a point $\boldsymbol{x}$ in $\partial\mathcal{M}$, there are an open neighborhood A of $\boldsymbol{x}$, an open neighborhood B of $\boldsymbol{0}$ in $\mathbb{R}^N$ and a diffeomorphism $\varphi : A \to B$ such that $\varphi(\boldsymbol{x}) = \boldsymbol{0}$ and

$$\varphi(A \cap \mathcal{M}) = \{y = (y_1, \ldots, y_N) \in B : y_{M+1} = \cdots = y_N = 0 \text{ and } y_M \geq 0\}.$$

Reasoning on the fact that the conditions (a) an (b) of the definition can not hold simultaneously for any point of $\mathcal{M}$, it is possible to prove that

$$\varphi(A \cap \partial\mathcal{M}) = \{y = (y_1, \ldots, y_N) \in B : y_M = y_{M+1} = \cdots = y_N = 0\}\,.$$

This completes the proof. □

Let us now see that, given a M-manifold $\mathcal{M}$, correspondingly to each of its point $\boldsymbol{x}$ it is possible to find a local M-parametrization.

Theorem C.3
For every $\boldsymbol{x} \in \mathcal{M}$, there is a neighborhood A' of $\boldsymbol{x}$ such that $A' \cap \mathcal{M}$ can be M-parametrized with a function $\sigma : I \to \mathbb{R}^N$, where I is a rectangle of $\mathbb{R}^M$ of the type

$$I = \begin{cases} [-\alpha, \alpha]^M & \text{if } \boldsymbol{x} \notin \partial\mathcal{M}\,, \\ [-\alpha, \alpha]^{M-1} \times [0, \alpha] & \text{if } \boldsymbol{x} \in \partial\mathcal{M}\,, \end{cases}$$

and $\sigma(\boldsymbol{0}) = \boldsymbol{x}$.

Proof
Consider the diffeomorphism $\varphi : A \to B$ given by the above definition and take an $\alpha > 0$ such that the rectangle $B' = [-\alpha, \alpha]^N$ be contained in B. Setting $A' = \varphi^{-1}(B')$, we have that A' is a neighborhood of $\boldsymbol{x}$ (indeed, the set $B'' =]-\alpha, \alpha[^N$ is open and hence also $A'' = \varphi^{-1}(B'')$ is open, and $\boldsymbol{x} \in A'' \subseteq A'$). We can then take the rectangle I as in the statement and define $\sigma(\boldsymbol{u}) = \varphi^{-1}(\boldsymbol{u}, \boldsymbol{0})$. It is readily seen that σ is injective and $\sigma(I) = A' \cap \mathcal{M}$. Moreover, $\varphi_{(1,\ldots,M)}(\sigma(\boldsymbol{u})) = \boldsymbol{u}$ for every $\boldsymbol{u} \in I$; hence, $\varphi'_{(1,\ldots,M)}(\sigma(\boldsymbol{u})) \cdot \sigma'(\boldsymbol{u})$ is the identity matrix, so that $\sigma'(\boldsymbol{u})$ has rank M, for every $\boldsymbol{u} \in I$. □

Remark In the proof we have seen that $\mathcal{M}$ can be covered by a family of open sets of the type A'', so that for each of them there is a local M-parametrization σ, defined on an open set containing I and injective there, such that $A'' \cap \mathcal{M} \subseteq \sigma(I)$. Restricting if necessary the sets A'', this property still holds if instead of A'' we take an open ball $B(\boldsymbol{x}, \rho_{\boldsymbol{x}})$. Moreover, if $\boldsymbol{x}$ is a point of the boundary $\partial\mathcal{M}$, the M-parametrization σ is such that the interior points of a single face of the rectangle I are sent on $\partial\mathcal{M}$.

We want to define an orientation for $\mathcal{M}$, which will automatically induce one also for $\partial\mathcal{M}$. Given $\boldsymbol{x} \in \mathcal{M}$, let $\sigma : I \to \mathbb{R}^N$ be a local M-parametrization, with $\sigma(\boldsymbol{0}) = \boldsymbol{x}$. Since $\sigma'(\boldsymbol{u})$ has rank M, for every $\boldsymbol{u} \in I$, we have that the vectors

$$\left[\frac{\partial\sigma}{\partial u_1}(\boldsymbol{u}), \ldots, \frac{\partial\sigma}{\partial u_M}(\boldsymbol{u})\right]$$

form a basis for a vector space of dimension M which will be called the **tangent space** to $\mathcal{M}$ at the point $\sigma(\boldsymbol{u})$ and will be denoted by $\mathcal{T}_{\sigma(\boldsymbol{u})}\mathcal{M}$ (in particular, if $\boldsymbol{u} = \boldsymbol{0}$, we have the tangent space $\mathcal{T}_{\boldsymbol{x}}\mathcal{M}$).

Now, once $\boldsymbol{u} \in I$ is considered, the point $\sigma(\boldsymbol{u})$ will belong also to the images of other M-parametrizations. There can be a $\tilde{\sigma} : J \to \mathbb{R}^N$ such that $\sigma(\boldsymbol{u}) = \tilde{\sigma}(\boldsymbol{v})$, for some $\boldsymbol{v} \in J$. We know from ▶ Chap. 3 how it is possible to change the orientation to such a $\tilde{\sigma}$ with a simple change of variable. Hence, we can choose these local M-parametrizations so that the bases of the tangent space $\mathcal{T}_{\sigma(\boldsymbol{u})}\mathcal{M} = \mathcal{T}_{\tilde{\sigma}(\boldsymbol{v})}\mathcal{M}$ associated to them all be coherently oriented; this means that the matrix which permits to pass from one basis to the other has a positive determinant. We will call **coherent** such a choice.

A coherent choice of the local M-parametrizations is therefore always possible, remaining in a neighborhood of $\boldsymbol{x}$. But we are interested in the possibility of making a *global* coherent choice, i.e., for *all* possible local M-parametrizations of $\mathcal{M}$. This is not always possible: for example, it can be seen that this can not be done for a Möbius strip, which is a 2-manifold.

Whenever *all* the local M-parametrizations of $\mathcal{M}$ can be chosen coherently, we say that $\mathcal{M}$ is **orientable**. From now on we will always assume that $\mathcal{M}$ is orientable and that a coherent choice of all the local M-parametrizations has been made. We then say that $\mathcal{M}$ has been **oriented**.

Once we have oriented $\mathcal{M}$, let us see how it is possible to define, from that, an orientation on $\partial\mathcal{M}$. Given $\boldsymbol{x} \in \partial\mathcal{M}$, let $\sigma : I \to \mathbb{R}^N$ be a local M-parametrization with $\sigma(0) = \boldsymbol{x}$; recall that in this case I is the rectangle $[-\alpha, \alpha]^{M-1} \times [0, \alpha]$. Being $\partial\mathcal{M}$ a $(M-1)$-manifold, the tangent vector space $\mathcal{T}_{\boldsymbol{x}}\partial\mathcal{M}$ has dimension $M-1$ and is a subspace of $\mathcal{T}_{\boldsymbol{x}}\mathcal{M}$, which has dimension M. Hence, there are two versors in $\mathcal{T}_{\boldsymbol{x}}\mathcal{M}$ which are orthogonal to $\mathcal{T}_{\boldsymbol{x}}\partial\mathcal{M}$. We denote by $\nu(\boldsymbol{x})$ the one which is obtained as a directional derivative $\frac{\partial\sigma}{\partial v}(0) = d\sigma(0)\boldsymbol{v}$, for some $\boldsymbol{v} = (v_1, \ldots, v_M)$ with $v_M < 0$. At this point, we choose a basis $[v^{(1)}(\boldsymbol{x}), \ldots, v^{(M-1)}(\boldsymbol{x})]$ in $\mathcal{T}_{\boldsymbol{x}}\partial\mathcal{M}$ such that $[\nu(\boldsymbol{x}), v^{(1)}(\boldsymbol{x}), \ldots, v^{(M-1)}(\boldsymbol{x})]$ be a basis of $\mathcal{T}_{\boldsymbol{x}}\mathcal{M}$ oriented coherently with the one already chosen in this space. Proceeding in this way for every $\boldsymbol{x}$, it can be seen that $\partial\mathcal{M}$ is thus oriented: we have assigned to it the **induced orientation** from that of $\mathcal{M}$.

Assume now that $\mathcal{M}$, besides being oriented, be **compact**. Given a M-differential form $\omega : U \to \Omega_M(\mathbb{R}^N)$, with U containing $\mathcal{M}$, we would like to define what we mean by integral of ω on $\mathcal{M}$.

In the case when $\omega_{|\mathcal{M}}$, the restriction of ω to the set $\mathcal{M}$, be zero outside the support of a single local M-parametrization $\sigma : I \to \mathbb{R}^N$, we simply set

$$\int_{\mathcal{M}} \omega = \int_{\sigma} \omega .$$

In general, we have seen that $\mathcal{M}$ can be covered by some open sets A'' of $\mathbb{R}^N$, which we can assume to be open balls, for each of which there is a local M-parametrization $\sigma : I \to \mathbb{R}^N$ with $A'' \cap \mathcal{M} \subseteq \sigma(I)$. Being $\mathcal{M}$ compact, there is a finite sub-covering: let it be given by $A''_1, \dots, A''_n$. The open set $V = A''_1 \cup \dots \cup A''_n$ contains then $\mathcal{M}$. We now need the following result.

Theorem C.4

There exist some functions $\phi_1, \dots, \phi_n : V \to \mathbb{R}$, of class $\mathcal{C}^\infty$, such that, for every $\boldsymbol{x}$ and every $k \in \{1, \dots, n\}$, the following properties hold:

(*i*) $0 \le \phi_k(\boldsymbol{x}) \le 1$,

(*ii*) $\boldsymbol{x} \notin A''_k \Rightarrow \phi_k(\boldsymbol{x}) = 0$,

and, for $\boldsymbol{x} \in \mathcal{M}$,

(*iii*) $\sum_{k=1}^n \phi_k(\boldsymbol{x}) = 1$.

Proof

Let $A''_k = B(x_k, \rho_k)$, with $k = 1, \dots, n$. Consider the function $f : \mathbb{R} \to \mathbb{R}$ defined by

$$f(u) = \begin{cases} \exp\left(\frac{1}{u^2-1}\right) & \text{if } |u| < 1 , \\ 0 & \text{if } |u| \ge 1 , \end{cases}$$

and set

$$\psi_k(\boldsymbol{x}) = f\left(\frac{\|\boldsymbol{x} - \boldsymbol{x}_k\|}{\rho_k}\right).$$

Then, for every $\boldsymbol{x} \in V$, it is $\psi_1(\boldsymbol{x}) + \dots + \psi_n(\boldsymbol{x}) > 0$, and we can define the functions

$$\phi_k(\boldsymbol{x}) = \frac{\psi_k(\boldsymbol{x})}{\psi_1(\boldsymbol{x}) + \dots + \psi_n(\boldsymbol{x})} .$$

The required properties are now easily verified. □

The functions $\phi_1, \dots, \phi_n$ are said to be a **partition of unity**. Since each $\phi_k \cdot \omega_{|\mathcal{M}}$ is zero outside the support of a single local M-parametrization, we can define the integral of ω on $\mathcal{M}$ in this way:

$$\int_{\mathcal{M}} \omega = \sum_{k=0}^{n} \int_{\mathcal{M}} \phi_k \cdot \omega .$$

It is possible to prove that such a definition does not depend neither on the (coherent) choice of the local M-parametrizations, nor on the particular partition of unity defined above.

We can now state the analogue of the **Stokes–Cartan theorem**.

Theorem C.5
If $\omega : U \to \Omega_M(\mathbb{R}^N)$ is a M-differential form of class $\mathcal{C}^1$ and $\mathcal{M}$ is a compact, oriented $(M+1)$-manifold contained in U, then

$$\int_{\mathcal{M}} d\omega = \int_{\partial\mathcal{M}} \omega$$

(provided the orientation on $\partial\mathcal{M}$ is the induced one).

Proof
Let us first assume that there is a local M-parametrization $\sigma : I \to \mathbb{R}^N$ such that

$$\sigma(I) \cap \partial\mathcal{M} = \emptyset ,$$

and $\omega_{|\mathcal{M}}$ is equal to zero outside $\sigma(I)$. By the injectivity of σ and the continuity of ω, we have that ω has to be zero on all points of the support of $\partial\sigma$, so that

$$\int_{\mathcal{M}} d\omega = \int_{\sigma} d\omega = \int_{\partial\sigma} \omega = 0 .$$

On the other hand, since ω is zero on $\partial\mathcal{M}$,

$$\int_{\partial\mathcal{M}} \omega = 0 .$$

Hence, the identity is verified in this case.

Assume now that there is a local M-parametrization $\sigma : I \to \mathbb{R}^N$ which sends the interior points of a single face I_j of I on the boundary of $\mathcal{M}$ and that $\omega_{|\mathcal{M}}$ is equal to zero outside $\sigma(I)$. Then,

$$\int_{\mathcal{M}} d\omega = \int_{\sigma} d\omega = \int_{\partial\sigma} \omega ,$$

and since ω is zero outside the support of $\partial\sigma$ except for the points coming from I_j, which belong to $\partial\mathcal{M}$, we have

$$\int_{\partial\sigma} \omega = \int_{\partial\mathcal{M}} \omega \,.$$

Hence, even in this case the identity holds.

Consider now the general case. With the above found partition of unity, each of the $\phi_k \cdot \omega$ is of one of the two kinds just considered. Being

$$\sum_{k=1}^{n} d\phi_k \wedge \omega = d\left(\sum_{k=1}^{n} \phi_k\right) \wedge \omega = d(1) \wedge \omega = 0\,,$$

we then have

$$\begin{aligned}
\int_{\mathcal{M}} d\omega &= \sum_{k=1}^{n} \int_{\mathcal{M}} \phi_k \cdot d\omega \\
&= \sum_{k=1}^{n} \int_{\mathcal{M}} d\phi_k \wedge \omega + \sum_{k=1}^{n} \int_{\mathcal{M}} \phi_k \cdot d\omega \\
&= \sum_{k=1}^{n} \int_{\mathcal{M}} d(\phi_k \cdot \omega) \\
&= \sum_{k=1}^{n} \int_{\partial\mathcal{M}} \phi_k \cdot \omega \\
&= \int_{\partial\mathcal{M}} \omega\,,
\end{aligned}$$

and the proof is completed. □

A final remark about orientation. In ▶ Chap. 1, when dealing with functions f defined on an interval $[a, b]$, the natural orientation of $\mathbb{R}$ was implicitly used, suggesting also the introduction of the notation $\int_b^a f = -\int_a^b f$.

In ▶ Chap. 2 this issue was not emphasized, even if some care was needed in the Change of Variables theorem, where only diffeomorphisms φ are allowed, and the factor $|\det \varphi'|$ appears in the main formula.

In ▶ Chap. 3 we did not really define an orientation for a M-surface σ, but nevertheless introduced the concept of *equivalent* M-surfaces which could have the *same* or the *opposite* orientation. Moreover, each M-surface σ induces an orientation on its boundary $\partial\sigma$, in such a way that the Stokes–Cartan theorem appears as the theory's most natural conclusion.

Finally, in this appendix, we have seen that the orientation of a differentiable manifold plays a crucial role. It is a delicate question, and it could be not so easy to verify in practice. But, again, the whole theory is motivated by the Stokes–Cartan theorem, which can be stated in this framework in its full elegance.

Appendix D
The Banach–Tarski Paradox

Alessandro Fonda

A. Fonda, *The Kurzweil-Henstock Integral for Undergraduates*,
Compact Textbooks in Mathematics,
https://doi.org/10.1007/978-3-319-95321-2

Let us start with the following

Definition D.1

Two subsets $\mathcal{A}, \mathcal{B}$ of $\mathbb{R}^3$ are said to be **equi-decomposable** if there are some sets $A_1, \dots, A_n$, which are pairwise disjoint, and $B_1, \dots, B_n$, also pairwise disjoint, such that

$$\mathcal{A} = A_1 \cup \dots \cup A_n \,, \qquad \mathcal{B} = B_1 \cup \dots \cup B_n \,,$$

and each B_i happens to be a roto-translation of A_i . In that case, we will write $\mathcal{A} \sim \mathcal{B}$.

We recall that a roto-translation is just the composition of a rotation with a translation. It is not difficult to prove that $\sim$ is an equivalence relation. Let us introduce the following notations:

$$B = \{x \in \mathbb{R}^3 : \|x\| \leq 1\} \,; \qquad S = \{x \in \mathbb{R}^3 : \|x\| = 1\} \,.$$

Moreover, once we have fixed a vector $v \in \mathbb{R}^3$ such that $\|v\| > 2$, for every subset E of $\mathbb{R}^3$ we denote by E_T the translation of E by the vector v :

$$E_T = \{x \in \mathbb{R}^3 : x - v \in E\} \,.$$

Let us state and prove the astonishing **Banach–Tarski theorem**.

Theorem D.2
It truly happens that $B \sim B \cup B_T$.

How is this possible? Everybody knows that a roto-translation of a body preserves its volume, while the Banach–Tarski theorem affirms that two sets may be equidecomposable even if they have different volumes. The point is that, among the "pieces" $A_1, \dots, A_n$, there clearly must be some which are non-measurable, hence there is no volume to be preserved! This is a serious task, which shows us the importance of the concept of *measurable set*.

Before starting the proof, we need an important property of the rotations in $\mathbb{R}^3$. We will say that two rotations ρ_1, ρ_2 of $\mathbb{R}^3$ are **independent** if, by a finite number of compositions of the elements in $\{\rho_1, \rho_1^{-1}, \rho_2, \rho_2^{-1}\}$, it is not possible to obtain the identity function, unless allowing the appearance of couples of the type $\rho_1\rho_1^{-1}$, $\rho_1^{-1}\rho_1$, $\rho_2\rho_2^{-1}$, $\rho_2^{-1}\rho_2$. We will call **simplified** the compositions where these couples are not allowed. In the following, we will consider only simplified compositions.

Lemma D.3
There exist two independent rotations of $\mathbb{R}^3$.

Proof
Let ρ be the rotation with angle $\arccos \frac{1}{3}$ in counter-clockwise direction around the x-axis and ϕ be the analogous rotation around the z-axis. We have that $\rho, \rho^{-1}, \phi, \phi^{-1}$ are represented by the following matrices:

$$\rho^{\pm 1} = \begin{pmatrix} 1 & 0 & 0 \\ 0 & \frac{1}{3} & \mp\frac{2\sqrt{2}}{3} \\ 0 & \pm\frac{2\sqrt{2}}{3} & \frac{1}{3} \end{pmatrix}, \qquad \phi^{\pm 1} = \begin{pmatrix} \frac{1}{3} & \mp\frac{2\sqrt{2}}{3} & 0 \\ \pm\frac{2\sqrt{2}}{3} & \frac{1}{3} & 0 \\ 0 & 0 & 1 \end{pmatrix}.$$

We will show that any simplified composition f of elements in $\{\rho, \rho^{-1}, \phi, \phi^{-1}\}$ can not be the identity, proving by induction the following proposition, for $n \geq 1$:
(P_n) *For every* f *having n components, one has that* $f(1, 0, 0)$ *is of the form*

$$\frac{1}{3^n}(a, b\sqrt{2}, c)\,,$$

where a, b, c *are integer numbers and* b *is not a multiple of 3.*
Without loss of generality, let us assume that the last component to the right of f be ϕ or ϕ^{-1}. If $n = 1$, we have $f = \phi^{\pm 1}$ and $\phi^{\pm 1}(1, 0, 0) = (\frac{1}{3}, \pm\frac{2\sqrt{2}}{3}, 0)$; hence, ($P_1$) holds with $a = 1$, $b = \pm 2$ and $c = 0$. Assume now that (P_k) holds for $k = 1, \dots, n$ and consider f having $n + 1$ components. Then, $f = \phi^{\pm 1} f'$ or $f = \rho^{\pm 1} f'$, where f' has n components

and $f'(1,0,0) = (a', b'\sqrt{2}, c')/3^n$, with b' not being a multiple of 3. In the first case,

$$f(1,0,0) = \phi^{\pm 1} f'(1,0,0) = (a' \mp 4b', (b' \pm 2a')\sqrt{2}, 3c')/3^{n+1},$$

while in the second case,

$$f(1,0,0) = \rho^{\pm 1} f'(1,0,0) = (3a', (b' \mp 2c')\sqrt{2}, c' \pm 4b')/3^{n+1}.$$

It still remains to prove that neither $(b' \pm 2a')$ nor $(b' \mp 2c')$ are multiples of 3. We will do it considering the four possible cases when f is one of the following:

$$\phi^{\pm 1}\rho^{\pm 1} f'', \quad \rho^{\pm 1}\phi^{\pm 1} f'', \quad \phi^{\pm 1}\phi^{\pm 1} f'', \quad \rho^{\pm 1}\rho^{\pm 1} f'',$$

where f'' does not appear at all if $n = 1$, while, if $n \geq 2$, f'' has $n - 1$ components and $f''(1,0,0) = (a'', b''\sqrt{2}, c'')/3^{n-1}$, with b'' not being a multiple of 3.

In the first case, $b = b' \pm 2a'$ and $a' = 3a''$, so that b is not a multiple of 3.

In the second case, $b = b' \mp 2c'$ and $c' = 3c''$, similarly as in the first case.

In the third case, $b = b' \pm 2a' = b' \pm 2(a'' \mp 4b'') = b' + (b'' \pm 2a'') - 9b'' = 2b' - 9b''$, hence b is not a multiple of 3.

In the fourth case, $b = b' \mp 2c' = b' \mp 2(c'' \pm 4b'') = b' + (b'' \mp 2c'') - 9b'' = 2b' - 9b''$, similarly as in the third case.

In any case, (P_{n+1}) holds, and the proof is thus completed. □

Proof of the Theorem

We denote by ρ, ϕ two independent rotations of $\mathbb{R}^3$, whose existence is guaranteed by Lemma D.3. Let $\mathcal{F}$ be the set of all the rotations which can be obtained as the composition of a finite number of elements in $\{\rho, \rho^{-1}, \phi, \phi^{-1}\}$, to which we add the identity. Let D be the subset of S made by the fixed points of the rotations in $\mathcal{F}$, except for the identity. Being $\mathcal{F}$ countable, Lemma D.3 tells us that D is countable, too. The sequel of the proof is divided into three steps.

Step 1 We want to prove that $S \setminus D \sim (S \setminus D) \cup (S \setminus D)_T$. Given $\boldsymbol{x} \in S \setminus D$, we define the orbit of $\boldsymbol{x}$ through $\mathcal{F}$:

$$\sigma(\boldsymbol{x}) = \{f(\boldsymbol{x}) : f \in \mathcal{F}\}.$$

It is easily seen that $\sigma(\boldsymbol{x})$ is contained in $S \setminus D$ and that two orbits either coincide or are disjoint. Hence, the set of all orbits makes up a partition of $S \setminus D$. By the axiom of choice, we can construct a set M taking a single point from each of these orbits.

Let us prove now that, varying $f \in \mathcal{F}$, the sets $f(M)$ generate a partition of $S \setminus D$. First observe that, taken $\boldsymbol{x} \in S \setminus D$, there is a $\boldsymbol{u} \in \sigma(\boldsymbol{x}) \cap M$; it will be $\boldsymbol{u} = g(\boldsymbol{x})$, for some $g \in \mathcal{F}$. Then, setting $f = g^{-1}$, we have that $f \in \mathcal{F}$ and $\boldsymbol{x} = f(\boldsymbol{u}) \in f(M)$. This proves that the sets $f(M)$ cover $S \setminus D$. Secondly, if there is a $\boldsymbol{x} \in f_1(M) \cap f_2(M)$, with $f_1, f_2 \in \mathcal{F}$, then both $f_1^{-1}(\boldsymbol{x})$ and $f_2^{-1}(\boldsymbol{x})$ belong to $\sigma(\boldsymbol{x}) \cap M$, and therefore coincide, by the way M has been defined. Then, $\boldsymbol{x} = f_2(f_1^{-1}(\boldsymbol{x}))$, which means that $\boldsymbol{x}$ is a fixed point for $f_2 f_1^{-1}$. Since $\boldsymbol{x} \notin D$, it has to be $f_1 = f_2$. this proves that the sets $f(M)$ are pairwise disjoint.

We now define the following sets:

$$A_1 = \bigcup \{f(M) : f \in \mathcal{F} \text{ starts on the left side with } \rho\},$$

$$A_2 = \bigcup \{f(M) : f \in \mathcal{F} \text{ starts on the left side with } \rho^{-1}\},$$

$$A_3 = \bigcup \{f(M) : f \in \mathcal{F} \text{ starts on the left side with } \phi\},$$

$$A_4 = \bigcup \{f(M) : f \in \mathcal{F} \text{ starts on the left side with } \phi^{-1}\}.$$

They form a partition of $(S \setminus D) \setminus M$, since the identity is excluded from those f starting on the left side with $\rho, \rho^{-1}, \phi, \phi^{-1}$. Let us prove that

$$A_1 \cup \rho(A_2) = S \setminus D, \qquad A_1 \cap \rho(A_2) = \emptyset,$$

$$A_3 \cup \phi(A_4) = S \setminus D, \qquad A_3 \cap \phi(A_4) = \emptyset.$$

If $\boldsymbol{x} \in S \setminus D$, it has to belong to one and only one of the sets $f(M)$, with $f \in \mathcal{F}$. If f begins with ρ, then $\boldsymbol{x} \in A_1$; otherwise, since we are only considering simplified compositions, $\rho^{-1} f$ begins with ρ^{-1}, so that $\rho^{-1}(\boldsymbol{x}) \in A_2$, that is $\boldsymbol{x} \in \rho(A_2)$. The two first equalities then follow. Analogously one proves the two second ones.

If we define $B_1 = A_1$, $B_2 = \rho(A_2)$, $B_3 = (A_3)_T$, $B_4 = (\phi(A_4))_T$, we thus have seen that

$$(S \setminus D) \setminus M \sim (S \setminus D) \cup (S \setminus D)_T .$$

Consider now the bijective function

$$\beta : (S \setminus D) \setminus M \to (S \setminus D) \cup (S \setminus D)_T ,$$

which is obtained using the respective roto-translations of the sets A_1, A_2, A_3 and A_4: we will have that β coincides with the identity on A_1, with the rotation ρ on A_2, with the translation by the vector $\boldsymbol{v}$ on A_3, with the rotation ϕ followed by the translation by $\boldsymbol{v}$ on A_4. We set $\alpha = \beta^{-1}$ and define the set

$$C = \bigcup_{n=0}^{\infty} \alpha^n((S \setminus D)_T) = (S \setminus D)_T \cup \alpha((S \setminus D)_T) \cup \alpha^2((S \setminus D)_T) \cup \dots ,$$

where α^n denotes the function α iterated n times.

We now show that

$$(S \setminus D) \setminus \alpha(C) = ((S \setminus D) \cup (S \setminus D)_T) \setminus C .$$

Indeed, if $\boldsymbol{x} \in (S \setminus D) \setminus \alpha(C)$, then $\boldsymbol{x}$ does not belong to

$$\alpha(C) = \alpha((S \setminus D)_T) \cup \alpha^2((S \setminus D)_T) \cup \dots ,$$

and since surely it does not belong to $(S \setminus D)_T$, then $\boldsymbol{x}$ is not an element of C. Consequently, $\boldsymbol{x} \in ((S \setminus D) \cup (S \setminus D)_T) \setminus C$. Vice versa, if $\boldsymbol{x} \in (S \setminus D) \cup (S \setminus D)_T \setminus C$, then, not being an element of C, $\boldsymbol{x}$ does not belong to $(S \setminus D)_T$, so that $\boldsymbol{x} \in S \setminus D$; moreover, since $\alpha(C) \subseteq C$, $\boldsymbol{x}$ can not belong to $\alpha(C)$, hence $\boldsymbol{x} \in (S \setminus D) \setminus \alpha(C)$.

In conclusion, being $C \sim \alpha(C)$, we have

$$S \setminus D = \alpha(C) \cup ((S \setminus D) \setminus \alpha(C)) \sim C \cup (((S \setminus D) \cup (S \setminus D)_T) \setminus C) = (S \setminus D) \cup (S \setminus D)_T ,$$

which is what we wanted to prove.

Step 2 We will now prove that $S \sim S \cup S_T$. Let ℓ be a line, passing through the origin $\mathbf{0}$ of $\mathbb{R}^3$, which does not intersect the countable set D. Let us see that there is a rotation ψ, having ℓ as axis, such that

$$\forall n \geq 1 \quad \psi^n(D) \cap D = \emptyset .$$

Consider first the set Θ_1 made of those angles which are rational multiples of 2π. These angles determine all the possible rotations f around the axis ℓ for which a point of D comes back to itself by applying an iterate f^n. Consider now two points $\boldsymbol{x}_1, \boldsymbol{x}_2 \in D$ which lie on the same orthogonal plane to ℓ : the point $\boldsymbol{x}_1$ will be moved onto the point $\boldsymbol{x}_2$ by a rotation around the axis ℓ by a certain angle $\theta_{12} \in [0, 2\pi[$. Define the set Θ_2 made of the angles of the type $(\theta_{12} + 2\pi m)/n$; these angles determine all the possible rotations f around the axis ℓ for which the point $\boldsymbol{x}_1$ is moved on $\boldsymbol{x}_2$ by applying an iterate f^n. Since D is countable, the sets Θ_n can be ordered in a sequence. Moreover, each of the sets Θ_n is countable and hence their union is countable, too. It is therefore sufficient to take an angle which does not depend on any Θ_n to find a rotation ψ with the required property.

Consider now the set

$$\widetilde{D} = \bigcup_{n=0}^{\infty} \psi^n(D) = D \cup \psi(D) \cup \psi^2(D) \cup \dots .$$

Then,

$$S = \widetilde{D} \cup (S \setminus \widetilde{D}) \sim \psi(\widetilde{D}) \cup (S \setminus \widetilde{D}) = S \setminus D .$$

Analogously one sees that $S_T \sim (S \setminus D)_T$. Recalling the equi-decomposability proved in Step 1, using the symmetry and the transitivity of the relation $\sim$, we obtain the equi-decomposability we are looking for.

Step 3 Associating to each $\boldsymbol{x} \in S$ the radius without its origin $\{\lambda \boldsymbol{x} : \lambda \in]0, 1]\}$, we can proceed exactly as above to prove that $B \setminus \{\mathbf{0}\} \sim (B \setminus \{\mathbf{0}\}) \cup (B \setminus \{\mathbf{0}\})_T$. Let now η be a rotation by an angle which is an irrational multiple of 2π around the line $\{(x, 0, 1/2) : x \in \mathbb{R}\}$. Consider the set $\tilde{O} = \{\eta^n(\mathbf{0}) : n \geq 0\}$. Then,

$$B = \tilde{O} \cup (B \setminus \tilde{O}) \sim \eta(\tilde{O}) \cup (B \setminus \tilde{O}) = B \setminus \{\mathbf{0}\} .$$

In an analogous way one sees that $B_T \sim (B \setminus \{\mathbf{0}\})_T$; by the symmetry and the transitivity of the relation $\sim$, we conclude that $B \sim B \cup B_T$, and the theorem is thus proved. □

As we said above, the Banach–Tarski theorem should make anyone aware of the strange consequences one can face when dealing with non-measurable sets. Those sets are constructed by the use of the axiom of choice. Nevertheless, this axiom is still usually accepted for its remarkable usefulness in modern mathematics.

Appendix E
A Brief Historical Note

Alessandro Fonda

A. Fonda, *The Kurzweil-Henstock Integral for Undergraduates*,
Compact Textbooks in Mathematics,
https://doi.org/10.1007/978-3-319-95321-2

In this appendix, I will take a brief look at the historical evolution of the concept of integral, without laying any claims to completeness. In particular, I would like to stress the role played by the Riemann sums, in different stages.

The primary motive for the integral calculus stems from geometrical problems such as the computation of the length of a curve, the area of a surface, planar or not, and the volume of a solid. Some of these were already computed since Ancient Greek times, in particular by Eudoxus (4th century B.C.) and Archimedes (3rd century B.C.). The method used at that time was based on two steps: first, a candidate for the integral was found by the use of approximations which could resemble the Riemann sums; then, the rigorous proof was given by the so-called "exhaustion method". Obviously, the notations were completely different from ours, and the procedure followed was mainly geometrical rather than analytical.

The main significant change in the setting of the problem was made in the seventeenth century, when Descartes discovered Analytic Geometry. In particular, differential calculus started to be developed as the method of determining the tangents to a given curve. A fundamental step was then made by Leibniz (1682) and Newton (1687). They independently understood the link between differential and integral calculus, finding that, if F is a function whose derivative coincides with f, then

$$\int_a^b f(x)\,dx = F(b) - F(a)\,.$$

This is what we have called the Fundamental Theorem.

Thanks to the contributions of Euler (1768) and others, the theory of primitivable functions was developed in the set of those functions defined by a single analytical formula or by a power series, which were about the only functions deemed worthy of interest at that time. In this framework, the above formula was sometimes used as the very definition of integral, and the Riemann sums approach took on a secondary role.

However, a fundamental research by Fourier (1811) put in evidence that such a theory was too restrictive: it was time to develop a theory which could deal with some discontinuous, not primitivable functions. Cauchy (1823) was the first to provide a rigorous basis for the theory, but it was Riemann (1854) who introduced the definition of integrable function, the one we have called R-integrable in this book, which we recall here.

Definition E.1

A function $f : [a, b] \to \mathbb{R}$ is R-integrable and its integral is some real number A if the following property holds: for every $\varepsilon > 0$ there is a real number $\delta > 0$ for which one has

$$\left| \sum_{j=1}^{m} f(x_j)(a_j - a_{j-1}) - A \right| \leq \varepsilon ,$$

for every choice of points a_j and x_j such that

$$a = a_0 < a_1 < \cdots < a_{m-1} < a_m = b ,$$

with

$$a_j - a_{j-1} \leq \delta , \quad \text{and} \quad a_{j-1} \leq x_j \leq a_j .$$

Nevertheless, the Cauchy-Riemann theory was not the final solution to the problem of integration: Volterra (1881) gave an example of a primitivable and bounded function which is not R-integrable on $[a, b]$. The main problem was that this procedure did not take into account the particular properties of the functions involved. According to Borel, it was like "a ready-made outfit which doesn't suit to each and everyone". The problem of finding a theory where both R-integrable and primitivable functions were integrable persisted.

Lebesgue (1902) faced this problem introducing the following integration procedure. Given $f : [a, b] \to \mathbb{R}$, bounded with non-negative values, and $C > 0$ such that $0 \leq f(x) < C$ for every $x \in [a, b]$, let us consider the subdivision of the interval $[0, C[$ (and not of the interval $[a, b]$!) in n parts $\left[\frac{j-1}{n}C, \frac{j}{n}C\right[$, and consider the sets

$$E_n^j = \left\{ x \in [a, b] : \frac{j-1}{n}C \leq f(x) < \frac{j}{n}C \right\} .$$

The problem is how to "measure" these sets. In general, given a set $E \subseteq [a, b]$, Lebesgue calls "outer measure" of E the infimum of the set of all sums $\sum_k (d_k - c_k)$, finite or countable, obtained by considering a covering of E with the intervals $[c_k, d_k]$.

Denoting by $\mu^*(E)$ this exterior measure, the "interior measure" of E is given by $\mu_*(E) = (b-a) - \mu^*([a,b] \setminus E)$. The set E is measurable if $\mu_*(E) = \mu^*(E)$, in which case this number is called the "measure of E" and is denoted by $\mu(E)$. At this point, Lebesgue defines the integrable functions, which in this book have been called L-integrable.

Definition E.2

A function $f : [a,b] \to \mathbb{R}$, such that $0 \le f(x) < C$ for every $x \in [a,b]$, is L-integrable and its integral is a real number A if the following property holds: for every $\varepsilon > 0$ there is a natural number $\bar{n}$ such that, taken $n \ge \bar{n}$, the sets E_n^j are measurable and

$$\left| \sum_{j=1}^{n} \frac{j}{n} C \mu(E_n^j) - A \right| \le \varepsilon .$$

The definition is then extended to functions which are not bounded and with arbitrary real values in the following way.

Definition E.3

Assume that the function $f : [a,b] \to \mathbb{R}$ has non-negative values but is not bounded above. In this case, f is said to be L-integrable on $[a,b]$ if, for every positive integer k, the function $f_k(x) = \min\{f(x), k\}$ is L-integrable on $[a,b]$ and the sequence of the integrals $(\int_a^b f_k)_k$ has a finite limit as k tends to $+\infty$. Such a limit is said to be the "integral of f on $[a,b]$". Whenever f not only has non-negative values, it is said to be L-integrable if both its positive part f^+ and its negative part f^- are L-integrable; the integral of f is then the difference of the respective integrals of f^+ and f^-.

Again according to Borel, this procedure "is custom made, and perfectly adapts to the properties of each function". Lebesgue's theory was completely satisfactory for several aspects. It was shown that every R-integrable function is also L-integrable and the two integrals have the same value, and if a function is primitivable and bounded, then it is L-integrable.

However, the problem of the integrability of not bounded primitivable functions remained unsolved at the time. It finally found a solution some years later by Denjoy (1912) and Perron (1914), who extended Lebesgue's theory with two different but equivalent approaches. While Denjoy used a transfinite induction method starting from Lebesgue definition, Perron's method is more in line with the formula for the integral of primitivable functions. Let us take a simplified and brief look at Perron's method.

We call *lower-primitive* of f a function F_- such that $F_-(a) = 0$ and, for every $x \in [a,b]$, it is $F_-'(x) \le f(x)$; similarly, we define an *upper-primitive* F_+ of f, by changing the inequality sign. We say that f is *almost primitivable* if there are a lower-primitive and an upper-primitive. In the following definition, we call P-integrable the functions which are integrable according to Perron.

Definition E.4

A function $f : [a, b] \to \mathbb{R}$ is P-integrable if it is almost primitivable and

$$\sup\{F_-(b) : F_- \text{ is a lower primitive }\} = \inf\{F_+(b) : F_+ \text{ is an upper primitive }\}.$$

In such a case, this value is called "integral of f on $[a, b]$".

While it is clear that every primitivable function is P-integrable, to prove that every L-integrable function is P-integrable (with the same value for the integral) is rather complicated. It comes out that a function f is L-integrable if and only if both f and $|f|$ are P-integrable.

The approach to the definition of integral had been set aside for a long time when Kurzweil (1957) and Henstock (1961) proposed the following modification to Riemann's definition: after observing that the two conditions $a_j - a_{j-1} \leq \delta$ and $a_{j-1} \leq x_j \leq a_j$ can be replaced by

$$x_j - \delta \leq a_{j-1} \leq x_j \leq a_j \leq x_j + \delta\,,$$

without modifying the definition, in order to adapt the procedure to each function they decided to allow greater freedom of choice for the δ. With a simple but far-reaching insight, the formerly constant δ could now vary in the interval $[a, b]$, according to the needs of the function. This is how they reached the definition that we have adopted in this book:

Definition E.5

A function $f : [a, b] \to \mathbb{R}$ is integrable (according to Kurzweil and Henstock) and its integral is a real number A if the following property holds: for every $\varepsilon > 0$ there is a function $\delta : [a, b] \to \mathbb{R}$, with positive values, for which one has

$$\left| \sum_{j=1}^{m} f(x_j)(a_j - a_{j-1}) - A \right| \leq \varepsilon\,,$$

for every choice of the points a_j and x_j in such a way that

$$a = a_0 < a_1 < \cdots < a_{m-1} < a_m = b\,,$$

and

$$x_j - \delta(x_j) \leq a_{j-1} \leq x_j \leq a_j \leq x_j + \delta(x_j)\,.$$

Surprisingly, a function is integrable (according to Kurzweil and Henstock) if and only if it is P-integrable, and in that case the value of the integral is the same.

Some years later Mac Shane (1969) proved that, by modifying in the above definition the condition $a_{j-1} \leq x_j \leq a_j$ with the less restrictive one $x_j \in [a, b]$, an alternative definition of L-integrable functions is obtained.

In conclusion, it can be said that the Riemann sums have played and still play a fundamental role in the theory of integration, even if with alternate fortunes. Intuitively used by the ancient Greeks, they were placed on the back burner once the theory of Leibniz and Newton was introduced, only to be back in the limelight thanks to Cauchy and Riemann. Overshadowed once again by the theories of Lebesgue, Denjoy and Perron, they proved to be important again in the work of Kurzweil and Henstock, who re-introduced them to unify these theories in an easy and intuitive way. And they are the subject of interesting research developments even today.

Bibliography

1. R.G. Bartle, A Modern Theory of Integration, American Mathematical Society, Providence, 2001.
2. Z. Buczolich, The g-integral is not rotation invariant, Real Analysis Exchange 18 (1992/93), 437–447.
3. A. Fonda, Lezioni sulla Teoria dell'Integrale, Ed. Goliardiche, Roma, 2001.
4. R.A. Gordon, The Integrals of Lebesgue, Denjoy, Perron, and Henstock, American Mathematical Society, Providence, 1994.
5. R. Henstock, Definitions of Riemann type of the variational integrals, Proceedings of the London Mathematical Society 11 (1961), 402–418.
6. R. Henstock, Theory of Integration, Butterworths, London, 1963.
7. R. Henstock, The Generalized Theory of Integration, Clarendon Press, Oxford, 1991.
8. J. Kurzweil, Generalized ordinary differential equations and continuous dependence on a parameter, Czechoslovak Mathematical Journal 7 (1957), 418–449.
9. J. Kurzweil, Nichtabsolut Konvergente Integrale, Teubner, Leipzig, 1980.
10. J. Kurzweil, Henstock–Kurzweil Integration: Its Relation to Topological Vector Spaces, World Scientific, Singapore, 2000.
11. S. Leader, The Kurzweil–Henstock Integral and its Differentials, Marcel Dekker, New York, 2001.
12. P.Y. Lee, Lanzhou Lectures on Henstock Integration, World Scientific, Singapore, 1989.
13. P.Y. Lee and R. Vyborny, The Integral. An Easy Approach after Kurzweil and Henstock, Cambridge University Press, Cambridge, 2000.
14. T.Y. Lee, Henstock–Kurzweil integration on Euclidean spaces, World Scientific, Singapore, 2011.
15. J. Mawhin, Analyse: Fondements, Techniques, Evolution, De Boeck, Bruxelles, 1979–1992.
16. R.M. McLeod, The Generalized Riemann Integral, Mathematical Association of America, Washington, 1980.
17. E.J. McShane, Unified Integration, Academic Press, New York, 1983.
18. W.F. Pfeffer, The Riemann Approach to Integration, Cambridge University Press, 1993.
19. W.F. Pfeffer, Derivation and Integration, Cambridge University Press, Cambridge, 2001.
20. M. Spivak, Calculus on Manifolds, Benjamin, Amsterdam, 1965.
21. Ch. Swartz, Introduction to Gauge Integrals, World Scientific, Singapore, 2001.

A. Fonda, *The Kurzweil-Henstock Integral for Undergraduates*,
Compact Textbooks in Mathematics,
https://doi.org/10.1007/978-3-319-95321-2

Index

A. Fonda, *The Kurzweil-Henstock Integral for Undergraduates*,
Compact Textbooks in Mathematics,
https://doi.org/10.1007/978-3-319-95321-2

Printed in the United States
By Bookmasters